MICROPSYCHOLOGY

THE ROOTS OF MENTAL FUNCTION AND SEXUALITY

HOW A FEW BRAIN MECHANISMS SHAPE THE MIND

Yehuda Salu

This book may be purchased for educational, business, or sales promotional use. For ordering information please contact: orders@MicroPsychology.org

Please visit the book's website: www.MicroPsychology.org

First printing 2008
Printed in the U.S.A.

ISBN   1 434 847 810

EAN-13   978 1 434 847 812

1. Psychology. 2. Artificial intelligence. 3. Human sexuality.

CONTENTS

PROLOG

Solomon's Bewilderment

שלשה המה נפלאו ממני וארבע לא ידעתים
דרך הנשר בשמים
דרך נחש עלי צור
דרך אניה בלב ים
ודרך גבר בעלמה

משלי ל ,י"ח-י"ט

Three things have bewildered me, and four I have not known:
The way of the eagle in the sky;
The way of a snake upon a rock;
The way of a ship in high seas;
And the way of a man with a young woman.

Proverbs 30, 18-19

King Solomon, "The Wisest of All Mankind", lived some three thousand years ago. In two of the books attributed to him, Proverbs and Ecclesiastes, he handed us down advice, based on his observations of human nature and the world. The third book, The Song of Songs, is a hymn to love. His words of wisdom enlightened and guided generations of readers. His love verses are sung and savored to these days.

By his own admission, Solomon could not understand four commonplace, yet amazing, phenomena: how an eagle soars in the sky; how a legless snake crawls on bare rock; how a heavy ship sails across the deep sea; and how a man engages a young woman.

The first three phenomena are from the physical world, the fourth is spiritual. By bundling them together, is he alluding that physical and spiritual activities are similar? That matter and mind are one?

In the time that has passed since Solomon's days, humanity has solved his first three problems; the fourth is still open. A growing number of thinkers and scientists now believe that all mental activities are driven by concrete, physical processes. Although the complete picture is not clear yet, inroads were made into the mysterious realm of the human mind. The following is an account of a journey on those inroads. It touches on some of the issues related to mind and matter that have been clarified by modern science, and acknowledges some of the challenges that still lie ahead.

Behind the Scenes

Deep down, Sol felt that his robot would win the "War of the Androids". His robot was smarter. While all the others would bump into a rival and then clobber it to a pulp, his robot had a sharp eye and a brain. A primitive brain as it was, it could spot a charging rival. Like a bullfighter, it would wait patiently until the very last moment, back off quickly, and as the rival would be rushing by, overwhelm it with a single, devastating side-blow.

When he built the robot, Sol thought of possible scenarios and asked himself "What would I do in a similar situation?" He implemented his insights in the robot's brain. During the competition, Sol was glowing with pride when his brain-child was doing exactly what he himself would have done under the same circumstances. Sol's robot won the competition.

Actually, Sol tended to believe that humans are just very sophisticated robots. He would argue that every human action or feeling is the result of physical brain mechanisms. If we knew all those basic brain mechanisms, we could build robots that mimic human behavior. They would learn by themselves how to act in any environment. They would relate to each other, and would have feelings, plans, and thoughts – just like human beings. By observing the behavior of such robots, it would be impossible to distinguish between them and human beings – but for what they are made of.

Sol knew that we are far from building humanlike robots. He was amazed that although he understood how the brain controls our movements, he knew so little about how the brain regulates our emotions and mental activities.

When he was younger, hanging out with the boys, the topic of attraction to girls came up. His friends kept saying how they admire nice faces, long hair, and breasts. When Sol said that what attracts him most in a girl is a nice behind, those who did not roll up their eyes gave him a look that said "weirdo!" Sol noticed it. "Is something wrong with me?" he thought. Quietly and methodically, he compared himself to friends and to people around him. To his relief, he discovered that behinds were high on the list of many other boys, even though they would not admit it. To his dismay, he also realized that skinny, aloof supermodels, which were idolized by almost everyone else, did not attract him at all. Instead, chubby, outgoing girls were his type. It then struck him that while he himself was attracted to women, some people were gay. "Was I born with what would attract me sexually?" he started to wonder. "And if I was not born with those preferences, how did they develop?"

Now a college student, Sol still could not find satisfying answers to those puzzling questions. With his engineering attitude, general solutions such as "it is hereditary", or "it is all based on early childhood experiences" were too vague. He needed concrete mechanisms that he could put his finger on. Mechanisms so lucid and so well understood that one day they could be implemented in humanlike robots.

When Sol found out that professor Morie was offering a class entitled "Micropsychology – Fundamental Mental Mechanisms", he signed up for it.

Four students showed up to the first meeting. Professor Morie said that the class is about fundamental brain mechanisms that serve as building blocks for a wide variety of behaviors. Those mechanisms will be introduced, and their relationships with

4

various mental processes will be described. He asked the students on which topics they would like him to focus.

Becky, a biologist, said that she would like to learn how biology fits into the wider scheme of things. What is the connection between nerve cells and the human mind? Which parts of our behavior are determined by the genes, and which parts are molded by experience?

Claire, an education major, was interested in the basic mechanisms that the brain uses in learning. She felt that teachers who knew those mechanisms could teach more effectively. They would know how to present material in ways that facilitate its digestion and absorption by the brain.

Bob, a criminologist, said that he is interested to find out why some illicit behaviors could be modified by external interventions, while others could not be changed.

Sol, a computer scientist, indicated that, in general, he would like to know "how the mind ticks". In particular, he is curious about the brain mechanisms that underlie sexual behavior.

After a frank and short discussion, they all agreed that they would like to find out more about all those topics, and especially about sexual drives. As Bob said, "We appreciate stuff that we can put to good use."

Professor Morie told them that they would have to submit a paper that summarizes the class. It should be accurate and informative, and should be written for the general public.

The four students stayed after the meeting to discuss their assignment. Claire said that the assignment is not simple, because she knows almost nothing about the related scientific fields.

Sol said that that should not be a problem, because between the four of them, they could cover most of those scientific topics. It would be just a matter of sharing responsibilities.

Bob said that his main concern is how to write a scientific paper that would be interesting to readers with different scientific inclinations. He added, "I have to confess, my interests in biology are minimal, compared to my interests in human behavior. I am afraid that readers like me, when they run into detailed explanations of biological processes, would just drop the paper once and for all. When I read a general science paper, I like to get the overall picture, and I don't care too much about irrelevant details. I think that our paper should avoid scientific minutia."

Becky responded, "I beg to differ. I am fascinated by what Bob calls biological minutia. To me that is the essence of real science. I think that our paper must include scientific details."

Sol muttered, "So, to make everyone happy, we would have to include scientific details, and not include them. Mission impossible?"

After a short pause Claire said. "I enjoyed reading the dialogs of Plato. He was a masterful presenter of elaborate ideas. He himself would convey the main themes, and side issues and clarifications were brought up in discussions between the participants. Maybe we should adopt that format. We could write the paper as a transcript of a symposium. The main themes would be presented by Professor Morie, and we would present side issues, minutia, and the likes. If we do that, readers like Bob, and I don't mean it in any negative sense, would be able to skip what they consider irrelevant. They won't have to throw the baby with the water, so to say."

The four weighed that idea, and agreed. At the end of the semester, after attending the classes, they met, compared their notes, and completed the assignment. The following is their paper.

CHAPTER ONE

OBSERVATIONS

THE STARTING POINT

PSYCHOLOGY AND MICROPSYCHOLOGY

Morie

Understanding how the brain works has been, and still is, one of the main challenges of science. Different scientific disciplines look at the problem from different perspectives. For example, neurobiologists treat nerve cells as fundamental units. They study how nerve cells function individually, and how they interact with each other. Biochemists and pharmacologists treat molecules as the fundamental units. They focus on chemical processes that underlie the function of the brain.

We will treat the brain as an information-handling system. The fundamental units of our treatment are concepts. We will try to understand how the brain forms concepts and how it handles them as it controls our activities.

When you think about a certain concept, it evokes in you thoughts about other associated concepts. Concepts in our brain are inter-connected, and they are organized as information structures. The brain uses fundamental mechanisms to learn new concepts and to incorporate them into information structures. It then uses other fundamental mechanisms to extract concepts from those information structures and to use them in its mental activities.

We will deal mainly with the fundamental mechanisms that the brain uses for handling bits of information and for generating components of behaviors. These are the building blocks of all the brain's mental operations. These fundamental mechanisms constitute what may be called "micropsychology", to distinguish them from the overall macro-behaviors, which are the main staple of psychology. The few fundamental mechanisms that constitute micropsychology are the building blocks of the numerous mental processes that the brain performs.

We will describe how we learn the first concepts in some of the main information structures, and how the brain uses them to execute macro-behaviors. We will also outline how the more complex information structures evolve, and how complex behaviors are assembled.

We will choose some common behaviors, analyze their information structures, and show how they are built by basic brain mechanisms. Our first example will be sexual behavior. We will also show how the fundamental mechanisms are involved in a variety of other behaviors and higher mental processes, such as thinking and intelligence. In all our discussions the emphasis will be on micropsychology; how innate bits of behavior regulate the acquisition and the merger of new information, and bring about the observed macro-behaviors and mental activities.

SEXUAL ATTRACTION AND AROUSAL

Morie

Sexual behavior is a central behavior, which is intermingled with other behaviors. It has been widely studied, but it still has many mysteries. We will start by discussing sexual attraction and arousal from a behavioral point of view. We will focus on the core of those macro-behaviors; the core from which all individual variations sprout. We then will talk about fundamental micro-mechanisms that may explain those observed macro-behaviors. We will also show how those same micro-mechanisms apply to a wide variety of other macro-behaviors.

Bob

What do you mean by "the core of sexual attraction and arousal"? Why not deal with sexual behavior in general?

Morie

Sexual behavior is too wide of an area. Sexual attraction and arousal are the first stages of any sexual behavior. First, one has to be attracted to a potential sexual partner. Then, sexual arousal keeps the process going. These are two distinct, albeit interrelated, tasks. They are akin to finding food and then eating and digesting it.

Even sexual attraction and arousal are very wide areas, and we will limit ourselves to their core; the most basic parts. In general, attraction and arousal are being built throughout life; we'll concentrate on their early stages. The fundamental mechanisms that underlie those stages are common to everyone.

After its early stages, sexual behavior varies significantly from person to person. Sexual drives affect behaviors that are not sexual in nature, and emotional needs that are not sexual affect sexual behavior. For example, there are many differences between what drives the attraction of a teenager to his next-door girlfriend and what drives the attraction of an aging politician to a prize supermodel. We will focus on the common elements of those behaviors.

The line that separates sexual attraction from sexual arousal is blurred. If you feel aroused by a person, you are also attracted to that person. If you are attracted to a person, you are also aroused. In our discussion of sexual attraction, we will deal mostly with the question of sexual orientation. In our discussion of sexual arousal, we will deal with factors, in addition to gender, that are involved in arousal.

We will limit our discussions to the mental aspects of attraction and arousal. We will consider well-established biological processes that have been observed in a variety of sexual and non-sexual circumstances. Sexual attraction and arousal manifest themselves in many ways. We will be looking for the most fundamental processes with which the more elaborated behaviors could be constructed.

ACTIVITY-PLANS

The brain controls how the body responds to external stimuli. A stimulus may elicit one response, or it may trigger a sequence of responses, each of which may depend on subsequent stimuli that are sensed by the brain. The brain controls such sequences of activities by activity-plans. Some of the activity-plans are genetic, some are learned, and some have genetic and learned components.

GENETIC ACTIVITY-PLANS

PHEROMONES: GENETIC AROUSAL MESSENGERS

Morie

Let us examine now some animal activity-plans that are involved in sexual activities. In general, the sexual process requires the cooperation of two individuals. The process has many steps, and coordination between the activities of the partners is crucial for its successful completion. During the process, the individuals communicate about the stage in which they are found or that they have just completed. This communicated information enables each partner to coordinate its own activities with those of its mate. Visual, auditory, tactile and chemical cues may be used in the communication. The actual modes of communication depend on the species and on the particular stage of the sexual process. After a cue has been received by an individual, it is processed and eventually prompts a response.

Many animals use pheromones to convey to their partners the sexual stage in which they are found. The release of a specific pheromone by one individual and the grasping of its significance by the mating partner is one of many orchestrated processes that take part in each of the partners. Pheromone communication is genetic. Differences between the genes of the males and the females of a species account for the production and release of a pheromone by one of the sexes and its detection and incorporation by the other.

Bob
What is the difference between hormones and pheromones?

Morie
Pheromones are compounds generated and released by one organism, that when received by another organism of the same species, can cause in it behavioral or physiological changes. Hormones are compounds that are released by an organism into itself, and affect its own behavioral or physiological processes.

Claire
Would you say that pheromone communication is an example of a

process that is "nature", as opposed to "nurture"?

Morie

Yes, because the production and release of pheromones by one mate and their detection and implementation by the second mate have been linked in a number of species to specific genes.

Claire

Are there genetic activity-plans of sexual processes in animals that involve pheromones as well as other cues?

Morie

Virgin females of certain fruit flies are attracted to a pheromone released by males. Sensing that pheromone changes the "state-of-mind" of the female from foraging to reproduction. The female follows that pheromone, and prefers it to fruit scents that she encounters. Once next to a male, other chemical, visual and tactile stimuli are exchanged and guide the mating.

Luring potential mates by sex pheromones is not restricted to insects. Birds, fishes, reptiles and mammals, such as female dogs and lionesses in heat, spread pheromones to attract males from great distances.

Bob

Are pheromones involved in the human sexual process?

Morie

Receptors of sex pheromones have been identified in many species. The vomeronasal organ is an organ in which sensing and preliminary processing of pheromones take place. It is found in a variety of vertebrates, including reptiles and mammals, and it has been implicated in sexual and social behaviors. In humans, though, the vomeronasal organ develops between seven and twenty-five weeks of gestation, and then it regresses and disappears completely, turning into cartilage. As of now, no pheromone has been implicated as a necessary cue in human courtship or mating.

C. ELEGANS: A GENETICALLY HARD-WIRED SYSTEM

Morie

Before we delve into human sexual processes, I'd like to tell you about another animal, the C. elegans. C. elegans (Caenorhabditis elegans), a member of the phylum Nematoda, is a one millimeter long worm; the size of a coma in a book. It lives in the soil and in rotting vegetation, feeding on microbes such as E-coli. Due to its relatively short life span (2-3 weeks), C. elegans is amenable to genetic studies. Its genome is thirty times smaller than the human genome, but still, about 35% of its genes have DNA sequences similar to humans (homologs). Many of the C. elegans' genes function like mammalian genes.

C. elegans has two sexes: male and hermaphrodite. It reproduces through male-hermaphrodite mating and by hermaphrodite self-insemination. The hermaphrodite has 959 cells, the male has slightly more; all are visible with a microscope. Out of those cells, about 300 are neurons, whose synapses have been mapped.

Bob

What are neurons and synapses?

Morie

Neurons are nerve cells. Neurons are connected to each other. Some neurons are connected also to motor units or to sensory units. A connection between two neurons or between a neuron and a motor unit is called a synapse.

Some of the C. elegans' neurons form sensory structures that mediate responses to taste, smell, touch, temperature and light (it has no eyes, though). Other neurons form a central ring and two neural tracts, with connections to the sensory structures and to muscular units. C. elegans has 81 muscle cells, a digestive system and a reproductive system.

The mating behavior of the male consists of a sequence of activities, which are coordinated by cues from his hermaphrodite counterpart. In mating, every male C. elegans executes the same exact activities. First, sensory units on his tail recognize her. He then aligns himself against her, crawls back on her until he reaches her end, and turns his body. His hook explores her and finds the vulva. In rhythmical up and down motion, his spicules tap the lips of her vulva until they respond and open up. His anal sphincter shuts off his anus and the spicules begin their penetration. Initially, uterine signals prevent the advancing spicules from transferring sperm. Only after they have reached the right position, a second uterine signal mediates ejaculation. The spicules remain inserted for another 25-30 seconds, while she rests motionless. After that, he deposits a plug in the uterus and retracts the spicules.

This activity-plan is apparently hardwired in the C. elegans' neural system. By studying mutant males, mutant genes that cause defects in specific steps of the mating process were identified. Correlations were found between mutant genes and defects in the animal's responses to contact, backing, turning, locating the vulva, and inserting the spicules.

Overall, the activity-plans of the male and the hermaphrodite consist of sequences of innate activities. Each activity has a trigger and a terminator. Cues to activities that require coordination between the mates are provided by the sexual partner. Those cues are chemical or mechanical. The order of the activities is maintained by units that act as a terminator to one activity and as a trigger to the following one. This design allows for skipping stages of the process, if a subsequent stage has already been triggered. For example, if the male happens to be near the vulva at the onset, the un-needed crawling back would not be triggered, and the mating process will go on with the hook sensing the vulva.

Sol

Amazing! Three hundred nerve cells can regulate all those operations, in addition to regulating all other activities such as moving and feeding. No computer can even come close to such performance!

Claire

And the genome has precise control over the development of that system. It's pure "nature"; no "nurture".

THE CHALLENGE

Bob

I understand that the sexual activity-plans in humans depend on both nature and nurture. What are the learned elements?

Morie

The learned elements that may vary significantly from person to person include identification of a potential mate (sexual attraction), becoming aroused by own activities and by activities of the mate, and arousing the mate. Parts of those elements may be learned through experiences that are not sexual in nature, even before puberty. There are elements that are learned through direct involvement in sexual activity, such as body positioning, coordination with the mate, and timing.

A main difference between the sexual behavior of C. elegans and that of humans is in the nature of the cues that the partners exchange in order to sustain the process. All the cues of the C. elegans are genetic: the genes build the body parts that synthesize and release the pheromone in one of the sexes and the parts that receive and identify it and then trigger the appropriate activity in the other sex. The entire process is genetic. In humans, on the other hand, some of the process is learned. In different geographical locations and in different historic times humans have been relying on different cues to be sexually attracted and aroused by each other. Those cues were learned by fundamental genetic mechanisms.

One of our challenges is to come up with gene-mediated learning mechanisms that are flexible and robust enough to guaranty the continuation of the human species. Those learning mechanisms should be able to pick up sexual attraction and arousal cues in any society, engrave them in the minds of individuals in those societies, and guide them in their sexual interactions. In our discussions, we will try to identify such genetic learning mechanisms.

LEARNED ACTIVITY-PLANS

IMPRINTING

Morie

Before we discuss the nurture component in human sexual

activity-plans, I would like to bring some examples from other species, where part of the communication during the sexual process is not innate. It has to be learned by the newborn through experiences in the outside world.

A potential partner has to be identified, before the sexual process could start. Many birds identify a partner by visual cues that they learn at a very early age. Chicks of herring gulls have to learn how to identify their kind among other gull species that nest in the same area. They learn it by imprinting in their memory visual patterns extracted from the heads of their parents, while being fed by them. Those extracted patterns turn out to be not of the beaks. The shapes of the beaks are not specific enough. The chicks imprint the color of the eyes and the color patterns in the fleshy regions around the eyes of their feeders. Those small details are species specific. When the chicks grow, they use those memory imprints to identify mates of their own species. Experiments have shown that these are not hard-wired recognition mechanisms; they are actually learned. Chicks that were fed by foster parents of another species, or by their own parents whose eyes were painted to look like those of another species, tried to mate with gulls of that other species when they grew up. It seems that the neural networks in the brains of the chicks are hardwired to imprint the eyes of whoever feeds them. The imprints are used later on in life, during the sexual process.

Chicks of mallard ducks are programmed to imprint the image of their mother so that they could follow her. (They would imprint as 'mother' anything that moves around them in a certain time window after they have hatched). Adult mallard males use that imprint to identify adult mallard females as potential sexual partners. Male mallard chicks that were brought up by researchers using decoys of male mallards, try to mate with other males when

17

they mature. Female mallards disregard the imprint of their mother when they look for a mate. Instead, they use innate preference to typical visual features of males. This is an example that an image that was learned at an early age to serve one purpose of the young serves a different purpose of the adult. It also illustrates that a cue that was learned early in life by both sexes, is used later in life by males in different way than it is used (or not used) by females.

Claire

You have used the words learning and imprinting. What is the difference between the two?

Morie

In both cases, the brain keeps a record of a certain pattern of stimuli, be it visual, auditory, chemical or of any other kind. In imprinting, this can happen only within a limited time window. If the same stimuli happen before or after the critical period, they will not be recorded. A cow licks her newborn right after birth, to imprint in her brain its specific smell. That will enable her to recognize her offspring from all the other calves. If the calf is taken away from her at birth and returned after a few days, she will not be able to imprint its smell anymore.

Some people use the term imprinting to indicate an exact recording of an event. We will use it to indicate a recording that can take place only within a certain time window. Such a recording fills a missing part of an incomplete activity-plan. After the missing part has been added, the completed activity-plan becomes operational.

Bob

So, is 'imprinting' just another way of saying "an old dog cannot learn new tricks"?

Morie

Organisms are sometimes pre-wired to imprint information that would be crucial for their existence. Activity-plans are already hardwired in their system, except for arbitrary but relevant external information that needs to occur and to be imprinted. Once that information is imprinted, that activity-plan is complete and it could be executed. For example, in the duckling's brain, the entire activity-plan is hardwired, except for the shape of whatever would be moving near-by after the hatching. When that information is collected and imprinted, the activity-plan is complete and operational. The term 'imprinting' is usually used in that context.

Claire

Do humans imprint?

Morie

Yes. There are many things that humans can learn only within certain time windows, and once such a period is missed, they won't be able to learn it anymore, or will learn it only to a lesser extent. For example, in order to learn a language, children have to be immersed in a language-speaking environment. Nowadays, many people move to countries where they have to switch to a second language. It was found that only children younger than approximately six years of age can learn to speak the second language without any noticeable accent. The vast majority of the adults cannot do it.

In a few traumatic cases, severely abused children grew up in almost complete social isolation. After their rescue years later, they could not regain normal language capacity.

Micropsychology

Bob

Do we use the same mechanisms as the ducks for learning our sexual arousal and attraction cues?

Morie

The essence of imprinting is that there exists an incomplete activity-plan in the brain. The duckling can focus his eyes on a target, can walk to the target and stay next to it, but the target itself is not defined. Therefore, this incomplete activity-plan is not yet functional. There is also an innate learning mechanism, whose role is to complete that activity plan. The learning mechanism is set to detect certain cues. The duck detects nearby objects that move. When those external cues occur, as mother duck moves nearby, the learning mechanism is triggered. Cues which are associated with the trigger, such as the shape of mother duck, are picked-up and fill the missing parts of the activity-plan. That completes the activity-plan and it becomes operational. From now on, the duckling would stay next to his mother, based on her visual cues.

Humans, too, have incomplete activity-plans, whose missing parts are filled by imprinting. Some of those activity-plans are used later in life for sexual attraction and arousal.

Bob

What is the content of the incomplete sexual activity-plans in humans?

Morie

The circuitry of the incomplete sexual-attraction activity-plan would set in motion the interest of the individual in a potential mate. For example, it will focus your eyes on the person that attracts you, and you will feel urges to get closer and to touch that person. The activity-plan would also notify other parts of the brain about the new situation, so that they could participate in planning your activities. What is missing and has to be learned are the cues that would identify the attraction target. That learned (imprinted) part would include cues about the features of the other person that would trigger the sexual attraction activity-plan, e.g. physical characteristics of males or females in that particular society.

Similarly, the circuitry of the incomplete sexual-arousal activity-plan would set in motion the physiological aspects of sexual arousal that cause, among other things, hormonal releases and the typical pleasant feeling experienced with the arousal. The learned (imprinted) part would include what features of the other person, or what interactions with that person would trigger the sexual arousal, e.g. flirtation, hugs, kisses.

As innate learning mechanisms fill-in the cues in the incomplete activity-plans of sexual attraction and arousal, other innate mechanisms regulate the physical development of the sex organs. At puberty, all those innate learning mechanisms complete their tasks; the activity-plans have the body parts that would carry out the sexual activities.

Bob

Could you elaborate on this point?

Morie

Consider an average baby boy that will grow and become a straight man. When he is born, his sexual system is immature. It consists of partially developed sex organs and an immature neural network that connects them to the brain. They all need further development in order to become functional at puberty. Even nerves that are fully developed at birth could not generate physical arousal because both the brain centers that initiate arousal and the sex organs that respond are not mature yet. The brain's triggering centers have to learn to identify the external stimuli that should trigger sexual response. They have to learn to identify females and to identify arousing signals that females would send. The genetic learning mechanisms imprint into the brain's sexual-activity-plans cues that are extracted from experience. At the same time, other genetic mechanisms develop the sex organs to their physical maturity. At puberty, the triggering circuits complete their learning and the organs their maturation. The whole system becomes operational. The young man would be attracted by women, and his body would respond accordingly.

Becky

I believe that it would be justified to assume that many advanced species have pre-wired, incomplete activity-plans, which are completed and become operational by what we have called here imprinting. In primitive species, such activity-plans are completely innate. For example, the feeding systems of worms do not require any learning. The cues of what is food are innately embedded in those systems. Once food is sensed, it gets to the mouth and from there it is pushed through the digestive system. There is no learned component to this entire activity-plan. On the other hand, our feeding system has a learned component. We don't have to taste food in order to identify it. We can identify it by its shape. This part of our feeding activity-plan is learned, or "nurture". Once we bring food to our mouth, the rest of the process is "nature".

What is still unclear to me is what triggers new imprinting into the incomplete sexual activity-plans in humans. In ducks, the imprinting is initially triggered by near-by movements and then, in males, it evolves into sexual attraction. In seagulls, feeding activities in the nest trigger the imprinting of cues that are used later for mate identification. What triggers the imprinting of cues into the sexual activity-plans of humans?

TRIGGERS OF IMPRINTING

Morie

You hit the nail right on the head. This is a central question in understanding the roots of human sexual behavior. The answer to that question is not known yet. However, by analyzing available information we may get closer to answering it.

In general, almost everyone is sexually attracted to someone and aroused by something. The specific details of the attraction and the arousal vary between individuals. Therefore, the triggers of the imprinting mechanisms have to be universal, whereas the cues that they imprint may depend on individual circumstances. That would explain both the universality of those processes and the variability of their individual details.

Becky

What do you mean by universal triggers?

Morie

Universal triggers are triggers that have not changed throughout human history and that are the same for everyone. The same triggers have imprinted sexual attraction and arousal in humankind, always and everywhere.

Becky

What could those triggers be?

Morie

While in ducks the imprinting is triggered by motion, in humans sexual imprinting may be triggered by emotion.

Motion around a duckling triggers the imprinting of the associated stimulus, which later becomes a cue that drives behaviors. In analogy, certain emotions in humans trigger the learning of associated stimuli, which later become cues that drive other behaviors, including sexual attraction and arousal.

The basic emotions of humans are universal; they are hardwired in every one of us. They include joy, sadness, satisfaction, fear, disgust, surprise and some more. Newborns have hardwired reflexes and facial gestures that express those emotions in a universal way. Those reflexes and facial expressions do not depend on sex, race, ethnicity, geographical location, and they have not changed throughout history. All babies will make the same universal grimace of disgust when they taste for the first time new food that they do not like. The neural circuitries that control these behaviors are innately hardwired in us. They are "nature". They take part in a variety of brain operations, the same way that motion detection and other innate sensory circuitries do.

When certain fundamental emotions are stirred in the child, their associated cues are imprinted into the incomplete innate sexual activity-plans. Those imprinted cues determine what would sexually attract and arouse that person later in life. The fundamental emotions that trigger the imprinting are universal; all human beings possess them. The imprinted information, though, is individualistic; it depends on the personal experiences associated with those emotions.

Claire

What specific human emotions could trigger imprinting of cues into the activity-plans of sexual attraction and arousal?

Morie

Sexual behavior is about certain specific interactions with another person. The beginning of such interactions may be characterized as guarded attractions. Approach a potential mate, or allow a potential mate to approach you, but be on the alert. When you start to feel safe, the innate parts of the sexual activity-plans kick in. The innate emotions that may serve as triggers of imprinting should reflect those feelings: First, a stranger makes you feel apprehensive, or at least puts you on guard. Then you respond by an action that makes you feel safer in the presence of that stranger. The feeling of apprehension may disappear completely, or its remnants may still linger together with your feeling-safe. When this sequence of emotions is invoked in a child, associated stimuli may be imprinted and become cues of sexual activity-plans. The imprinted cues may represent an object, such as body parts of the stranger, or actions of the child, such as smiling to the stranger.

With time and experiences, the original imprinted cues are expanded and generalized. For example, if originally smiling to the stranger was the action that caused the transition from feeling apprehensive to feeling safe, saying something friendly to the stranger may be added as an optional activity to smiling.

Expansion of imprinted cues is a general process that the brain employs. For example, babies like to get attention. They first may realize that they get attention by crying. The crying is then expanded to include many other behaviors such as: behave mischievously, behave nice as expected, explicitly demand attention, or pretend that you don't care about getting attention. All these expansions are learned through experiences.

Claire

At what age can children feel the sequence of emotions of apprehension of strangers and its transition into feeling safe

25

around them?

Morie

At around six month of age, babies normally become attached to their mothers (or to their main caregiver). While up to this age they would smile at any person, at this juncture they reserve their genuine smiles only to people that are close to them. When approached by strangers, they would cling to their mothers for safety. Usually, it is not hard to tell that they enjoy watching the stranger from the safety of their mother. This combination of emotions is a universal behavior, which crosses societal and geographical boundaries. The babies then become aware of their own fear of strangers and realize that they can do things to abate it and feel safe again. They develop and test strategies for various interaction modes with other people, including ways of avoiding them. Those experiences become distinct concepts in their minds, and they could serve as an important component in learning sexual arousal.

Claire

How strong is that fear of strangers?

Morie

It varies in its intensity and in its duration. It is rarely very intense, or terrifying. It can be described as feeling intimidated by the presence of larger beings, and playing it safe. Often, it subsides a few minutes after evaluating the new situation. However, in some cases it evolves to complete avoidance of adults for prolonged periods of time.

With time, the baby realizes that his mother, too, is sometimes a source of tension, especially when she responds to his own activities. There are activities that she approves of, and activities that she disapproves. The baby likes her attitude towards him when she approves his activity, and dislikes it when rebuked. He learns that doing what she likes results in his gratification. This concept is then generalized in the baby's mind by replacing his mother with a generally significant person. Seeking satisfaction by doing something that pleases another person becomes a central concept in his mind, and it affects his activities throughout his life. It also becomes a component in sexual arousal.

To his mother's dismay, quite often the baby finds out that he can avoid her antagonism and the negative feelings that follow, by controlling her. If he adopts certain assertive or offensive stances towards her, she will not respond negatively to activities of his that she does not like. The generalization of the concept that controlling people suppresses their antagonism becomes an important notion throughout his life, including in his sexual arousal.

The brain processes information in parallel, and it can handle several emotions at the same time. For example, a baby may feel intimidated by a stranger, and at the same time feel safe as he clings to his mother. The perceived threat is hovering above him together with the security that he feels with his mother. This superposition of conflicting universal emotions may serve later as an ingredient in sexual arousal.

Bob

You have mentioned that in gulls, feeding activates the imprinting of associated visual patterns that happen to be found within the chick's field of view. Those visual patterns serve later-on as cues in mate identification. Similarly, does breast-feeding in humans cause the imprinting of cues, which serve later on as triggers of sexual activity-plans?

27

Morie

I doubt it for two reasons. First, nowadays, large numbers of babies are bottle-fed by male and female caregivers. If feeding activities were serving as imprinting triggers for features that are in the field of view of the nursing baby, large numbers of adults would be sexually aroused by and attracted to feeding bottles, or to someone holding such a bottle. I am not aware of that. Second, the visual patterns of mature gulls are universal and do not change with time. Therefore, patterns that are imprinted by the chicks remain valid as the chicks mature. This is not the case for humans. At least in Western civilizations, visual cues that can be extracted from early caregivers are usually different from cues that would be available later in life for identifying a mate and becoming sexually attracted and aroused. Just consider how fashion and gender appearances change in one's lifetime.

Feeding activities in humans serve as triggers of learning the concept of 'person'. This concept starts to crystallize around stimuli that originate from the caregiver, such as the facial features of the mother, features of her voice, footsteps, smell, softness, and so on. The concept 'person' is then expanded through more experiences. It sprouts many other related concepts such as child, grown-up, mother, father, etc. Those concepts take part in other concepts including cues in sexual activity-plans. So, I believe that breast feeding in humans, and feeding in general, helps to form and shape the concept of person. Breasts and chests become sexual cues due to their involvement in situations that make the child feel safer in the presence of threats.

SPROUTING OF CONCEPTS

Claire

I understand that new concepts evolve from existing ones. What are the mechanisms that regulate this process?

Morie

New concepts sprout from established ones by several mechanisms. One mechanism is by replacing a component of the established concept by another element. Another is by eliminating part of the root concept. Let's look at few examples.

At an early age, a child develops the concept 'in order to feel safe with mother, I have to please her'. This is an established root concept, from which other concepts sprout.

By eliminating the beginning and replacing 'mother' with 'everyone' the concept 'I have to please everyone' sprouts from that root concept.

New sprouted concepts are put to test and evaluated. Based on testing it in real life situations, the concept 'I have to please everyone' may be reinforced, eliminated, or modified to something like 'I have to please significant people'.

The concept 'in order to feel safe with a significant person I have to please him', and some of the concepts that sprout from it are crucial to the social development, to sexual function, and to the self-appreciation of the individual. When combined with other sexual cues, it serves as a trigger to the sensation of sexual gratification. It creates that sensation in those who are aroused by pleasing their partner.

With time, the repertoire of activities for pleasing a sexual partner expands. The specifics of those activities depend on the environment in which the individual has been raised. Those activities fall into two major categories. One category includes activities that create positive emotions in the sexual partner, such as joy, relaxation, curiosity and safety.

Another major category is activities that eliminate negative emotions, such as apprehension, that the partners might feel towards each other. We will see that feeling safe with another person is the common denominator of many sexual arousers.

Claire

Why do you suggest that learning sexual cues evolves over time from non-sexual experiences?

Morie

For two reasons. First, look around you. Part of our sexual attraction depends on non-sexual cues such as eye color, hair color, skin color, smile, height, weight, length of legs, temperament, and so on. Those cues can be learned, at least at the beginning, before concepts about gender have been acquired by the child.

Second, learning is a progressive process. It starts by learning general concepts, which are then refined, partitioned and elaborated. 'People' is one of the first concepts that the baby will learn. Then, she will become attracted to some people more than to others. The cues to the attractive people will be classified in her mind according to the kind of emotion with which they are associated. Only then, cues that relate to gender, which is a concept that crystallizes later in her mind, would be added to the mix.

STATES-OF-MIND

Morie

Most animals are sexually active only during certain time windows. Humans, on the other hand, can engage in sex year round. However, it appears that humans and animals have two distinct states-of-mind: sexual and non-sexual. All animals relate to the world around them according to their prevailing state-of-mind: Solitary animals seek the company of mates and foraging animals ignore food when they are in sexual state-of-mind. Human perceptions of the world and of their internal feelings depend too on whether sex is involved or not.

In animals, the transition from one of those states-of-mind to the other is regulated by hormones. Hormonal levels, in turn, may be influenced by external conditions such as weather or food availability. In humans, hormones affect the overall level of the libido. In addition, neural processes, conscious and non-conscious, determine if a situation is treated sexually or non-sexually.

Activity-plans control the behavior of an individual in response to ongoing external conditions. Sexual activity-plans may result in behaviors different from those guided by non-sexual activity-plans, even if they are responses to the same prevailing conditions. The state-of-mind in which an organism is found determines which activity-plans would actually be carried out.

LEARNING SEXUAL AROUSAL

FEELING SWAP

Bob

You said that cues that contribute to sexual arousal and to the identification of sexual mates evolve from individual, early, non-sexual, emotional experiences. Those emotions are combinations of feeling safe with another person while feeling somewhat threatened. The learned cues are imprinted into innate, incomplete activity-plans, and become fixated. They determine what would sexually-arouse that person later in life.

Morie

Right.

Bob

If feeling safe with another person in the presence of some threat is a central element in sexual arousal, why not give it its own name? We could use the acronym feeling SWAP, for feeling Safe With Another Person. It could stand also for feeling Safe With A Partner.

Morie

Good idea. Let's do it. So, feeling SWAP in a child may trigger the imprinting of sexual cues. Feeling SWAP in an adult is an element of the trigger of sexual arousal.

Becky

So, feeling SWAP has a dual role: (1) feeling SWAP triggers the imprinting of new cues into the arousal activity-plan, and (2) feeling SWAP triggers arousal. That is what happens in conditioning, as was demonstrated for the first time by Pavlov. In Pavlov's famous experiment, a bell sounded when food was given to a dog. After several repetitions, the dog salivated to the sound of the bell, even if food was not given. Sensing the food caused two things: salivation, and expanding the repertoire of stimuli that cause salivation (to include also the bell's sound). Feeling SWAP is the analog of sensing the food. Feeling SWAP causes two things: expanding the repertoire of things that cause sexual arousal, and becoming aroused.

Morie

Right.

SEXUAL SWAP and SOCIAL SWAP

Claire

Why we do not become sexually aroused when we are children

and we feel SWAP?

Morie

Because the rest of the system is not physically mature yet.

Claire

If so, why we do not become sexually aroused, as adults, when we do something that pleases another significant person and we feel SWAP?

Bob

Let me support that question by an example. In a bar, if I offer a drink to a lady that I do not know, and she accepts, I feel the beginning of sexual arousal. If, on the other hand, I offer a drink to a guy that I do not know, and he accepts, I just feel friendly, but not sexually aroused.

Morie

Did you have the same motivation when you offered the drink to the lady and to the guy?

Bob

Probably not.

Morie

The brain considered it, and treated the two gestures differently. When you offered the lady a drink, sexual thoughts were already brewing in your mind. The sexual state-of-mind began to take over, and activated the sexual arousal activity-plan. With the guy, you were in a different state-of-mind, and a different activity-plan was in effect.

The brain has two groups of activity-plans for feeling SWAP. One group is used in social context and the other in sexual context.

The goal of the social activity-plans is to enable social interactions between individuals. When performing those activity-plans, the individuals feel safe in their social environment, and that facilitates their interactions.

The goal of the sexual activity-plans is to create in the individuals sexual arousal, thus enabling them to engage in sexual activities.

Some activities, such as giving presents, create SWAP feelings in both social and sexual contexts. However, as we will see later, some activities create SWAP feelings in the sexual context and unsafe feelings in the social context and vice versa.

OBSERVATIONS IN OTHER SPECIES

Bob

What evidence is available to support the assertions about the role of feeling SWAP in sexual arousal?

Morie

There is plenty of circumstantial evidence.

Bob

I always prefer direct evidence - a smoking gun.

Morie

Oh, I am sorry. Smoking Gun could not come today. He has a coughing attack... Just kidding. We'll have to rely mainly on circumstantial evidence. But as you know, good circumstantial evidence is admissible and sufficient for reaching conclusions.

Bob

That's right.

Morie

One body of evidence comes from analyzing a variety of human behaviors during the sexual process. The second body of evidence comes from observing other animals. The fact that other biological systems operate under those rules makes it more plausible that Mother Nature has implemented them also in us humans.

Bob

Wow, wow, hold it! What do you mean Mother Nature? Why not Our Father in Heaven?

Claire

What's wrong with Mother Nature?

Sol

Relax, this is only a figure of speech. I am not a theologian, but my understanding is that God is neither male nor female. There is no 'male' unless there is 'female'. From your reactions, I gather that you believe in one Supreme Being. So, you cannot attribute to it sexual features. By doing so, you may imply that there are two Supreme Beings – God and his or her Partner.

Morie

Let's stay on course. What I mean is that there are rules according to which things happen in nature. Since humans are part of nature, some rules and mechanisms that apply to other species may also apply to humans, even if we cannot verify them directly on humans. Let's look first at examples from animals, and see if we can find traces of emotions that are involved in their sexual behavior.

The sexual process in many species includes a period of courtship. After a prospective mate has been identified, the partners forge a bond and arouse each other in courtship that, if successful, leads to mating. Courtships may be simple or elaborate. Some species exhibit behaviors with striking similarities to humans, while others are a class of their own. We'll look closer at some of the similar ones.

In many species, courtship intertwines gestures of greeting and aggression. To a human observer it appears that the partners are involved in an enjoyable game of controlled aggression and friendly closeness. They exchange playful gestures of fear, aggressiveness, relaxation, dominance and submission, borrowed from earlier periods of their lives, when they were fed and groomed by their parents and socialized with their siblings and peers.

Adult male grizzly bears lead solitary life most of the year, avoiding females and being avoided by them. They start their two months mating season with a search for a female. Once he comes across a female's track, the suitor will follow it, sniffing her day-beds and urine traces. She will run away from him and he will follow her, keeping his distance. If he gets too close too soon, while she is still afraid of him, she will charge at him aggressively and may smack him with her paw. The bigger and more dangerous male will retract and not retaliate. After a while, she gets used to the menace, and allows him to get closer. After they have made contact, they will nuzzle and chew on each other's head and neck and enjoy friendly wrestling bouts, like little cubs. There have been reports of males caressing and fondling females. After copulation, they separate and go their own separate ways.

The female in a few primates is the one who courts the male. When she is ready, she raises her tail and presents her hindquarters to the male. She looks at him backwards in a subordinate posture, making sounds that are a combination of cooing and fear. If he is aroused, he will clasp her ankles firmly with his feet, grab her buttocks with his hands, and copulate.

Bull elephants start their courtship by following a ready female. She may try to escape and he would follow her persistently. If she accepts, they will stroke each other with their trunks, intertwine them, and place their tips in each other's mouth, expressing mutual confidence and care.

Those examples illustrate how some species use the message: "you should not be afraid of me" to form bonds and to arouse sexual partners. Other species use the emotions of fear and safety in a different way. A male will direct his aggressive behavior towards other males, chasing them away from his only sweetheart, as in the case of Canada geese, or from a whole harem, as in the cases of elephant seals, zebras, deer, lions, and others. The implied message to the females in these cases might be "I am the strongest. With me you are safe". If that is not enough, explicit threats and direct attacks on straying females are initiated by the male, as in the case of fur seals.

Claire

Is feeling SWAP the only emotion that triggers sexual arousal in animals?

Morie

It is hard to tell precisely what emotions are experienced by animals during their courtships. It would be prudent to suggest that emotions other than fear and safety are aroused in many species. Some water birds have amazing courtship dances that might create sheer enjoyment in their mates. Similar effects may be achieved by stunning appearances, like those of the peacock or the greater bird of paradise. Nightingales and other songbirds use peaceful and enjoyable songs to build interest and arousal in their partners. The fundamental emotions behind those experiences could be the enjoyment of novelty and the diversion from the mundane. Some species, such as the crested fern, offer fish or other foods to their wooed. Here the fundamental emotion of pleasing through feeding is certainly in action. And the list goes on. One thing is clear: fundamental emotions, which play a role also in non-sexual experiences, take part in the sexual process, especially in mate identification and in arousal.

OBSERVATIONS IN HUMANS

OFF-MAINSTREAM

Morie

Let us now focus on the second body of evidence, which supports the assertion that sexual cues in humans are built around fundamental emotions, which are first experienced in a non-sexual context. Paramount are the combined emotions of feeling safe while fear is in the background (feeling SWAP).

Many couples enjoy role acting that arouses them sexually. Each partner plays a role that appears to be non-sexual, but the basic emotions created by the play trigger sexual arousal. The general scheme of such plays is the interaction between two individuals, one dominant and the other submissive. In some cases, those plays are very subtle, and in others, they are obvious and unmistakable. The latter are referred to as sadomasochism (SM), bondage and discipline (BD), and discipline and submission (DS). The players are referred to as dom and sub or top and bottom. Both players play their roles willingly and full heartedly, identifying with their roles. Common characters in such plays include a nurse and a patient, a guard and a prisoner, a teacher and a student, a master and a slave, a trainer and a dog, a parent and a child, a goddess and a worshiper, a human and a footstool, and more. The fundamental emotion of the top in all those plays can be characterized as feeling SWAP with a controlled partner. The emotion of the bottom is feeling SWAP by pleasing a significant person. Those two basic emotions are present in humans since infancy.

The fact that any of the roles can be played by a male or by a female supports the assertion that fundamental cues for sexual arousal are acquired through non-sexual experiences in infancy, when the concept of gender is not fully formed in the mind of the infant.

Claire

Are physical pain and humiliation involved in those plays?

Bob

It is up to the participants to choose how dominance and submission are expressed. They may choose benign symbols and gestures to express and arouse their emotions, but they may also choose harsh means. It depends on the participants. If both enjoy and want it, physical and mental pains are involved. If it is not done between adults with total free consent and desire of all parties, it is not considered sexual play; it may be a crime.

Claire

It *may be* a crime or it *is* a crime?

Bob

The way things are, it is not always a crime. In the workplace, a boss may humiliate a subordinate, who does not want to be treated like that. The subordinate accepts it for realistic reasons, but definitely does not derive any pleasure from it. The humiliation and suffering of the subordinate may arouse the boss sexually. If the humiliation is tied only to work performance, and no external signs of the boss's sexual arousal are indicated, it would not be considered a crime.

Claire

Is SM a mental disease?

Becky

It is a controversial issue. According to the Diagnostic and Statistical Manual of Mental Disorders (DSM-IV tr), the main diagnostic reference in the USA, to be considered a clinical disorder, these fantasies, urges, or behaviors must occur for a significant period of time and must interfere with either satisfactory sexual relations or everyday functioning. There is also a sense of distress within these individuals. The International Classification of Diseases (ICD-10), which is adopted by the World Health Organization, classifies such behavior-patterns as disorders of clinical significance.

Bob

Why use those behaviors as examples for making our point?

Morie

Because in those cases we do not rely on what people choose to tell us about their emotions and thoughts. We draw conclusions from observed activities that are being practiced by millions of people. Our premise is that certain non-sexual emotional experiences in early life, especially those related to SWAP, provide the cues that later play a part in sexual arousal, healthy or not. Moreover, there is a strong body of evidence that people who are not engaged in that kind of role-playing still use those same motifs in their sexual fantasies. They have the same thoughts, but they do not carry them out. They arouse themselves, as sole practitioners or as partners, using those same cues.

Claire

Why do some people resent those sexual-role-plays?

Bob

As members of society, individuals have to know how to engage with their peers, their superiors and their inferiors. Those who violate the rules of engagement risk themselves. In sexual-role-playing, players may fill roles that are inconsistent with their real social status, or do things that they would not do in real-life situations. For example, people high on the social ladder may play the most humiliating roles, or gentle souls may play the role of brutal slave drivers.

People who resent those sexual-role-plays believe, in one way or another, that there is a strong link between the sexual-plays and real life. Like moviegoers who do not separate the true from the screen personality of a movie star, some resent people because of their roles in sexual-plays. They rail against others and would hate to find such traits in themselves. Some may have concerns, which are sometimes founded, that the role-playing may overflow, and interfere with real-life activities.

Claire

It seems that in SM, the pleasure comes from enjoyment of inflicting physical and emotional pains, or from suffering pain and humiliation, and not from feeling SWAP.

Morie

SM stimulates many emotions, two of which are controlling a submissive and pleasing a dominant. These lead to feeling SWAP, which leads to sexual arousal. In general, the effects of other emotions may enhance or reduce the sexual arousal, and may have other consequences. Still, the cues of sexual arousal due to feeling SWAP are triggered in SM. A dom may experience several opposing emotions such as: I am safe through controlling the sub by inflicting pain and humiliation, I am good because the sub is enjoying the pain and the humiliation that I bestow. I am mean because I enjoy being cruel. The sub, too, may experience several opposing emotions: I am safe with the dom, the dom appreciates my pain and humiliation. I am hurting, and it is bad to be hurting. The brains of the players add up their emotions, and the totals turn out to be enjoyment. Had the total in any player been unfavorable, that player would not engage in that activity.

INTIMACY

Claire

Is feeling SWAP the only emotion that directly triggers sexual arousal?

Morie

It is not the only one, but it is a major one and its traces may be found in many situations that involve other emotions. It is quite common that when several emotions are at play, feeling SWAP is the one that most probably triggers the arousal. Here are some examples.

The natural tendency of humans and many animals to be intimate while engaging in sex might be another indication to that. Intimacy is characterized by retreating from the rest of the world, physically or mentally. The partners are seeking to be shielded from outside interferences, and the unwanted or unsafe consequences that may follow.

In addition to seeking a shielding from the outside, the partners are engaged in a variety of activities that strengthen their feeling SWAP. Activities that build trust, such as exchanging personal histories and secrets, are one example. Sharing a secret with someone is entrusting that person with a power to harm you. It is a way of saying indirectly: you should not be afraid of me because you have power over me. I am entrusting you with that power. On the other end, keeping another person's secrets is a way of saying: with me, you are safe; I am protecting your interests. I appreciate your trust in me.

Another example is activities that build up good mood. Sense-of-humor is very high on the list of the desired qualities in a partner. Generosity is another. Having a common interest is also at the top of the list. The good mood that is created by those activities drives away feelings of apprehension that a partner might have, thus increasing the feeling of safety.

THE RELATION BETWEEN SHAME, GUILT AND FEELING SWAP

Sol

Is SWAP a fundamental concept that serves as a core around which other concepts develop? We have seen that the brain builds new concepts from existing ones. For example, 'child', 'adult', 'man', 'woman', etc. are based on the core concept 'person'.

Morie

I believe so. Fear is an innate emotion, and the need to reduce it is an innate activity-plan. When the cause of fear is another person, the result is the need to feel SWAP.

'The cause of fear' is a concept that expands with experiences. A baby may be afraid of strangers because they are big. The concept 'fear of strangers' expands and it eventually includes not only physical properties of other people, such as their size, but also what those people can do or not do to us. A child fears being left alone by her mother, or being disciplined by a parent. The next developmental stage is that the child is afraid of doing something that will bring about such reactions from her parents. 'The cause of fear' has thus expanded to include certain activities that the child herself can do. The fear of doing such acts is at the basis of the pervasive emotions of shame and guilt.

Sol

Are you saying that the needs to reduce shame and guilt are expansions of the fundamental need to reduce fear?

Morie

Yes.

Sol

Thus, reducing shame and guilt reduces fear, which in turn creates the feeling of SWAP, and therefore may cause arousal, if the person is in a sexual state-of-mind.

Morie

That is a logical conclusion.

Bob

Any proofs?

Morie

Many normal women report that they are sexually aroused by

fantasies of being brutally raped. Those women would not seek and would not enjoy rough sex, and definitely would not enjoy being raped. They invoke those fantasies, and enjoy them, while having gentle sex.

Psychologists attribute those fantasies to unconscious thought-processes whose aim is to relieve shame and guilt feelings. Imagining being punished by a potent figure causes them to feel that they appease that figure, so that he would not be a threat to them. This is feeling SWAP. That relieves the unconscious feelings of shame and guilt and creates sexual arousal.

Men also enjoy similar arousal fantasies that relieve unconscious guilt and shame. It would be fair to consider those processes as extensions of SWAP.

FANTASIZING

Becky

One of the strategies that the brain uses to learn how to function in the environment is based on rewards and consequences. It strengthens behaviors that result in positive feelings and weakens behaviors that result in negative feelings. For example, we learn to eat tasty foods and avoid bland ones. We learn to look for shade when the sun is scorching, to open a window when the air is stuffy, and so on. You have pointed out that sexual pleasures could be guiding us in learning behaviors that satisfy us sexually. However, those pleasures are felt after puberty. Is there any sexual-like pleasure before puberty that can guide learning of sexual cues?

Morie

I believe that there is such a pleasure even before puberty, but that emotion is not fully formed. It is a precursor, which evolves to the feeling of sexual pleasure in the adult. The circuitries of sexual feelings and sensations together with the sex organs are all parts of one evolving system. They are all present from early on in an immature form, and they all mature at puberty.

A child enjoys fantasizing about a variety of subjects. It is fascinating to observe the enjoyment that children show as they use their imagination while playing with toys. Children fantasize about moving cars and crying dolls. They are thrilled to hear scary stories and to empathize with the characters. In their fantasies, children weave their emotions with thoughts about real world situations. Children build their emotional infrastructure as they grow up. There is no apparent reason why a child could not be able to fantasize about feeling safe with an intimidating person, and enjoy those thoughts. That enjoyment is achieved through circuitry that delivers some kind of a positive feeling. Before puberty, that positive feeling cannot be associated with full-scale sexual activity, because such activity is non-existent. It is associated mainly with thoughts. After puberty, that circuitry triggers the emotion of sexual pleasure. At that stage, the emotion can be associated with mature sexual activities of the body.

By thinking about sex, adults develop their perceptions about it. In their thoughts, they construct general scenarios in which they would like to be, and make plans for what they would do. The feedback provided by the sexual pleasure circuitry guides them in devising those plans. Later, they may carry out the plans. Similarly, the pre-puberty sexual pleasure circuitry must be guiding children in the development of their initial sexual pleasure programs. They develop all kinds of plans for what to do, in order to feel SWAP, and enjoy it. After puberty, those programs are incorporated with concrete sexual activities.

Claire

Do animals need to develop their sexual plans before they reach sexual maturity?

Morie

Copulation in a number of species, including the gorilla and the bonobo, is done in front of the entire troop. The young watch it with great interest and internalize it. In several species, the young rehearse sexual plays with peers and adults, of both the same and the opposite sex. It seems that these early practices are necessary for proper functioning in adulthood. Monkeys that have been raised in social isolation fail to breed as adults.

SWAP AND OTHER EMOTIONS

Morie

Based on the general principles of the brain's function, it is possible for sexual arousal to bypass the triggers of the SWAP cues. While those cues are learned in childhood, and they take part in early sexual experiences, the enjoyment of sexual activity after puberty may sway the brain to seek sex for the sake of its erotic physical pleasures. After experiencing sex and enjoying it, one might seek to repeat those experiences for the sheer physical sensations that they create. Wanting or expecting those pleasures would arouse the person, replacing the need to feel SWAP.

Sol

Mother Nature did not have to rely on the emotion of feeling SWAP as the guide for learning sexual arousal. She could have relied on partners making random contacts, just bumping into each other, and having sex. Then, like in Pavlov's conditioning, the partners would continue to seek sex, because of the rewarding pleasures that came with it.

Micropsychology

Claire

If so, some individuals would never experience sex, especially those who do not allow strangers to come too close to them.

Morie

One of the cases in which Mother Nature has employed the idea of random encounters is in plants. Pollen particles are spread by the wind and fertilize flowers that they encounter by chance. Mother Nature has many tricks up her sleeve, and for humans she has chosen a specific package that includes feeling SWAP.

Claire

One would think that after a bond has been established between a couple, based on feeling SWAP, that bond would last forever, unless something happens to shatter the safe feeling. However, we see many cases that couples are not arousing each other anymore, even if they still feel safe with each other. How come?

Morie

The emotion of feeling SWAP is most effective as an arouser if it is accompanied in the background by a sense of uncertainty about it. It is also more effective if there is a sense of transition from uncertainty to safety, like in courting. (Will he ask me out? Will she call me back? Oh, he did! Oh, she did! I am in heaven!). Since such enjoyable tension may be missing in couples that have known each other for a long time, the system could not rely solely on them. Through conditioning, the system provides for the development of other arousers that may replace the feeling SWAP. Many couples have known each other as young people, and aroused each other by their youthful displays. After many years, they keep arousing each other, even though those displays are all gone. Through conditioning and other innate mechanisms, their systems have developed and implemented other cues for arousal. Those cues are directly related to sexual activity, and they are not based on feeling SWAP. If such replacements are not formed, the fire may die down.

There is another major reason why some couples do not arouse each other anymore. One or both partners still have strong early programs of how to feel SWAP. Those programs, which may be conscious or unconscious, are based on perceptions that do not consider all the existing conditions. They rely on activity-plans that are unacceptable or controversial in mature conscious life. Thus, they lead to stress and dissatisfaction.

A very common example is of married men who want oral sex performed on them. To many, this is a symbol of male's dominance and female's submissiveness. A man that has the urge to feel SWAP through such actions may feel also conscious emotional conflicts: "With the same lips my wife will kiss our pure children? Since I cannot have it with my wife, she is not good enough." The wife may feel "I am not going to humiliate myself by giving him that. He is not considering my feelings. He is no good".

Claire

In other words, sexual incompatibility is caused by conflicting activity-plans that try to achieve the feeling of SWAP.

Morie

Sexual incompatibility has a variety of causes. Many cases arise because activities that trigger the feeling of SWAP, which stimulates sexual arousal, trigger also conscious emotions of shame and guilt. The combination of shame, guilt, and the sexual gratification of feeling SWAP may derail sexual enjoyment and compatibility.

Sol

How did this come about?

Morie

First, simple emotions that are aroused in interactions between the child and adults mold the patterns that will be used later in life as sexual arousers. Then, the time window for learning and updating basic sexual arousal cues is closed. Imprinted cues for sexual arousal cannot be erased anymore. They serve as the core around which new sexual activity-plans and concepts would be developed. All new plans and concepts of sexual arousal comply with the fundamentals of the fixated core.

In order to function in society, the child has to develop skills of controlling others and of being controlled. The emotion that goes hand in hand with those skills is feeling SWAP. After the fixation of the sexual core, the child keeps learning and refining activity-plans that will be used to create the feeling of SWAP in social interactions. Some social activity-plans contain fixated sexual activity-plans plus additional concepts. New emotions, such as shame and guilt, which were not available earlier, are combined with the sexual core and form new social activity-plans.

One example of a concept that serves as a component in social activity-plans is 'it is shameful to degrade yourself in order to feel SWAP'. Another example is 'it is shameful to degrade others so that you feel SWAP'. Later in life, when degrading activities are called upon to create sexual arousal, they may trigger also the emotions of shame and guilt. This creates an emotional conflict, which may be the cause of the incompatibility.

Bob

Is this the only type of conflicts between emotions that are related to sexual activities?

Morie

In general, stress is generated whenever a behavior or a thought evoke conflicting emotions. Initially, sexual arousal is linked to SWAP, but later in life, other emotions may evolve around it. In such cases, other emotions may be involved in conflicts related to sexual activities.

However, many conflicts are due to the fixation of certain cues as triggers of feeling SWAP, while at the same time those same cues can become triggers to other new developing emotions.

Sol

If that's the case, sex would be stressful for almost everyone. If the triggers of sexual arousal are fixated, whereas societal perceptions continue to evolve, conflicts between the two are inevitable. Still, sex for many people is not stressful, in spite of all the conflicts between their sexual and societal perceptions. How could this be explained?

Morie

Consider a fruit fly. The most sensible behavior for the female, at least for her own survival, would be to forage all the time for guavas and other delicious fruits. Yet, when the pheromones kick in, she abandons that behavior and follows her lust, so to speak. I do not think that she feels any inner conflicts. The reason is that she has shifted to a different state-of-mind; the sexual state-of-mind. What she does is consistent within that state-of-mind. Even if she had the neural circuitry for feeling conflicts, she would not feel them. She would be doing the only thing that her brain could come up with. The non-sexual (foraging) state-of-mind is completely shut-off when she is looking for a mate, thus no conflict is created.

When humans become sexually aroused, they also shift from a non-sexual to a sexual state-of-mind. However, the non-sexual state-of-mind is not completely shut off. It still functions in the background. The relative strength of those two co-existing states-of-mind affects the intensity of the conflict. The conflicts may be benign or severe. In a benign conflict, the activities of the individual affect only the emotions of that individual. In a severe conflict, those activities are in conflict with emotions of other members of the society, and may have legal ramifications. Those activities may affect the well-being of the individual and of others.

Like with any emotional stress, exposing its roots is a first step to recovery. In benign cases, a person that understands his or her situation and is able to separate between emotions that belong to the sexual domain and emotions that belong to the non-sexual domain may alleviate the stress or even eliminate it. However, if this does not help, and in cases of severe conflicts, professional intervention is needed.

Bob

Could you give an example of how exposing the roots of benign conflicts may help in resolving the stress that they generate?

Morie

There are people that become sexually aroused by sucking on the big toe of their partner. Imagine that you are one of them. You probably would not know how or why you have developed that urge, but you enjoy it. You might have developed it as a baby, when the toe of a caregiver became a cue for a person with authority. It then became a fixated part of the feeling SWAP circuitry in your brain.

Now, there are several options. If you and your partner understand and accept that that is how you express your SWAP feelings, there is no conflict and no stress. If you cannot accept it, try to find another activity that arouses you. If your partner makes fun of you, and that stresses you, try to enlighten her so that she understands human nature and changes her attitude. If she cannot change her attitude and you don't want to try something else, find another partner.

Bob

Assuming that I have a big toe fetish, and I don't want to expose myself. What can I do?

Morie

Use your imagination, and keep trying. Suck on something else which is more acceptable, and imagine that it is her toe. She might explode when you suck her earlobe. But don't tell her what is in your mind…

Bob

It seems that Mother Nature condones "mental cheating" and "faking" when it comes to sexual arousal.

Morie

It sure seems so. The arousal of each partner depends on the interpretation that he or she assigns to the ongoing activities. To arouse your partner, you have to create a setting that satisfies the partner's needs to feel SWAP. Your own arousal also depends on whether the ongoing activities make you feel SWAP. For both parties to be aroused, their activities have to induce SWAP feelings in both of them. However, in their minds, the partners are free to pursue their happiness, even by fantasizing.

SWAP AND PHYSICAL AROUSERS

Becky

We have limited our discussions to mental arousal. We have intentionally excluded physical activities, such as caressing, kissing, rubbing, and so on from our discussions. Are such activities purely physical, or do they have a mental dimension too?

Morie

They definitely have a mental dimension. Kissing, hugging, caressing, and in general touching involve invasion into the partner's domain. The actions themselves are considered tokens of friendship that the active partner conveys and the passive partner accepts and reciprocates. The partners are expressing with those actions their SWAP feelings towards each other.

Consider, for example, the one-year-old Jayjay. In his oral period, Jayjay learned the world by bringing it to his mouth. To his parents, it seemed that they had a sucking machine, and that their main job was to protect him from sucking hazardous objects. Jayjay would suck on the nose of anyone holding him. But now, it all changed. He agrees to be held by strangers, like aunt Bertha, but she wants more. She sweet-talks him into kissing her. Jayjay distinguishes between himself and others, and he appreciates his own domain. He does not like, or even feels threatened, if someone invades his domain. For him, having aunt Bertha's face on his lips is a threatening invasion of privacy. He listens to her cajoling and weighs his odds. By kissing her, her face will invade his domain, and may even cause an unpleasant taste on his lips. On the other hand, she may make some of those encouraging yells that he likes. He dares and kisses her, and everyone applauds. He feels gratification and is relaxed. He has just experienced an event of feeling SWAP. His brain has recorded the association between kissing and feeling SWAP. In the future, kissing may invoke in his brain that associated feeling, which trigger sexual arousal.

Becky

And what about intercourse positions; do they carry SWAP messages as well?

Morie

A variety of intercourse positions have been used and enjoyed since early times, and are described in many classical texts such as the Kama Sutra, and in modern literature. Some positions may be more efficient in eliciting stronger physiological stimuli than others, depending on the anatomical makeup and the agility of the partners. However, one cannot ignore the implied messages that the extra planning and coordination of some of the stylish positions send: What I am doing is going to increase the pleasure of my partner; my partner is trying to increase my pleasure. Therefore, I feel SWAP with my partner, my partner feels SWAP with me.

Bob

Talking about positions, isn't it interesting that Mother Nature has provided many species with only one intercourse position; male above female? Is Mother Nature implying that males are dominant?

Morie

I am glad that you brought it up. It is not only top and bottom; it is also front and back. In many species, the male mounts the female from her hind side. The hind side is more vulnerable than the front, where the teeth are. If you want to interpret the body language of the animals, you might say that the female is saying: I am exposing to you my vulnerable, non-combative side. I feel SWAP with you. The male from his perspective also feels SWAP because of that.

To me, it is more about willful cooperation than about coercion. It demonstrates that Mother Nature has designed the entire system so that arousal is associated with feeling SWAP. The other day I saw in the park a male mini-poodle trying to mount a pit bull bitch in heat. She did not appear submissive nor did he appear domineering. They were just trying their best to achieve a common goal in a harmonious way. If they were in another state-of-mind, he probably would not have dared to approach her.

Becky

Aren't you stereotyping the pit bull?

Morie

Oops!

Bob

Let's look at the so called "missionary position" (man on top, woman underneath while facing each other). Don't you think that this configuration symbolizes dominance of men and submissiveness of women?

Claire

Let me ask you a question. In society, there are people who work hard and people who relax and enjoy the hard work of others. Who is the dominant; the servant that sweats and waves the fan or the king that sits relaxed and enjoys the breeze?

Bob

The king that sits relaxed, of course.

Claire

And in the missionary position; who lies back, and who is huffing and puffing and working his tail off?

Bob

Touché.

SEXUAL ORIENTATION (ATTRACTION)

Claire

So far, we have discussed mainly sexual arousal. How about mate selection?

Morie

Mate selection activity-plans process two kinds of information: first, the gender of the selected mate, thus guiding sexual orientation. Second, other properties of the selected mate, thus restricting the attraction to subgroups within the desired sex. There are some overlaps between mate selection and sexual arousal. Selecting a potential mate cannot be completely detached from arousal, and vice versa. However, many features are unique to mate selection activity-plans.

The activity-plans for mate selection, like those of sexual arousal, are partially innate. They become operational after innate learning mechanisms have filled-in necessary parts, which have to be learned, bases on the actual experiences of the individual.

Sol

Do hormones affect the sexual orientation of an individual?

GENES AND HORMONES

Morie

From fertilization to about six weeks, the only difference between what will be a male or a female is in the chromosomes. Cells of female embryos have two X chromosomes and cells of male embryos have one X and one Y chromosome. The gonads are not yet differentiated. At six weeks, the differentiation between the sexes begins. The Y chromosome directs the gonads to become testes. If a Y chromosome is not present, the gonads develop into ovaries. Hormones secreted by the differentiated gonads regulate the development of several organs, including internal sex organs (e.g. ovaries, prostate), external sex organs (penis, vulva), and some neural pathways in the brain. Irregularities in the sex chromosomes may affect the development of any of those structures. After birth, males have greater concentrations of male hormones (androgens, e.g. testosterone) and females have greater concentrations of female hormones (e.g. progesterone, estrogens). However, each sex has both male and female hormone, only in different proportions. Around puberty, the sex hormones regulate the maturation of the sex organs and the development of the secondary sex characteristics (e.g. facial hair, voice, breasts, fat distribution, muscle mass). The hypothalamus begins regulating the production of sex hormones. In females, it regulates the cyclic secretion of sex hormones that control the female menstrual cycle. In males, it regulates the relatively continuous production of male sex hormones.

DEVELOPMENTAL DISORDERS

During prenatal development, sex hormones regulate the development of the internal and the external sex organs. In some disorders, there is an inconsistency between the developed internal and external sex organs. A newborn may have a fully or partially developed penis (morphologic male) with non-functional or no internal male sex organs. This happens if at a certain critical stage, a reduced activity of male hormones or increased activity of female hormones occurred, depending on the disorder. Similar prenatal processes occur in morphologic females, which may have a normal looking vulva or an enlarged clitoris. Usually, those children are raised according to their morphologic sex. However, their hormonal levels are determined by their internal sex organs. For example, a child with external male's sex organs is raised and treated as a boy, but his hormone levels are those of a girl. By studying such cases, researches could compare and contrast the role that hormones play in determining sexual orientation, versus the role played by upbringing and social interactions. It was found that there is no one-to-one correlation between the prevailing sex hormones and sexual orientation.

The populations of those studies were small, and the number of variables that may affect the behavior is large, so generalizations to the general population may be unreliable.

HORMONES AND BEHAVIOR

In adult men, lower levels of testosterone are linked to reduced libido. It may also have similar effects in women. Lowering the level of testosterone in male sex offenders by castration or by administration of estrogen or other synthetic steroids reduces their sex drive, or makes it physiologically impossible for them to have sex, but generally, it does not affect of whatever is left of their sexual orientation.

In men, higher testosterone levels were found in perpetrators of violent crimes and in subjects with impulsive behaviors. In general, testosterone influences aggressive and dominant behaviors, but usually it is not the sole or a necessary cause of those behaviors. There is evidence that in male competitions testosterone levels increase in the winner and decrease in the loser. That includes animals that fight for access to mates, human athletes, and fans of competing sports teams.

No significant differences in the concentration of testosterone or other sex hormones have been found between homosexual and heterosexual humans or between homosexual and heterosexual animals. Attempts to change the sexual preference of gay men by the administration of various sex hormones have failed.

The general roles of sex hormones may also be inferred from the case of the "opposite-sex identical twins". At the age of seven months, in a botched surgical procedure, the penis of one twin was severed and could not be reattached. By the advice of the doctors, it was decided to treat the injured twin as a girl. His testicles were removed, thus eliminating the main source of his male hormones. He underwent reconstructive surgery to make his genitals look feminine. He was not told about the mishap, and was treated and raised as a girl. During puberty, he was administered estrogens, which caused the development of feminine secondary sex characteristics (breasts, hair patterns). However, he always felt that something was not right, and as an adult, he could not function as a woman. After he has found the truth, he underwent another reconstructive surgery, this time to a man. He married a woman, adopted a child and felt whole with himself.

All these suggest that hormones affect the anatomy and gross features of behavior, but not the finer details of sexual orientation and arousal.

BRAIN MORPHOLOGY

Claire

Are there differences between the brains of males, females, heterosexuals, homosexuals, and bisexuals?

Morie

There are some differences between the brains of men and women. The brain consists of interconnected right and left hemispheres. Some activities are localized mainly in one hemisphere. For example, the language centers are found in the left hemisphere, whereas spatial processing of information takes place in the right hemisphere. When comparing the development of the hemispheres in the two sexes, the left hemisphere is more developed in females, and the right in males. Those differences are consistent with females, on the average, having the edge in linguistic skills, and males, on the average, having an edge in spatial orientation skills. This is a simplistic view, because the averages are taken over large populations. Quite often, individual differences go the other way.

Some centers in the brain show morphological differences that correlate with the sex of the individual and with sexual orientation. Four groups of cells, collectively called Interstitial Nuclei of the Anterior Hypothalamus (INAH 1 through 4), and some other cell groups show such differences. INAH 3 was shown to be twice as large in male heterosexuals as in male homosexuals. The exact role of those centers in sexual behavior is not yet known.

Claire

How big are those centers?

Micropsychology

Morie

Not big, about the size of a grain of sugar.

Bob

Like a few grains of sugar in the entire head? Isn't that a little too small to significantly affect those major behaviors?

Morie

Who knows? Remember the C. elegans? There, some 300 cells regulate the entire reproduction process. Here, we are talking about tens of thousands of cells. Anyhow, it is not clear if those centers are involved in planning and executing sexual activities, or if they are passive outcomes of those activities.

PARENTING

Bob

Is it true that parents serve as role models based on which children develop their sexual orientation: Children of heterosexual parents become homosexual because of "mistakes or bad examples" of their parents, and children in homosexual households are influenced by their homosexual parents?

Morie

Parents serve as role models for many behaviors, and their behavior may contribute to the sexual orientation of their children. However, it does not appear that children copy their sexual orientation from their parents, and that homosexuality of children is the result of "mistakes" or implied guidance of their parents. There is no evidence that being gay is caused by a dominant mother or a submissive father, or that being lesbian is caused by any particular types of parents. Nowadays, tens of millions of children grow up in single-parent heterosexual families and in families with one or two lesbian, gay, or bisexual parents. Scientific studies that were done failed to show any difference in the percentage of homosexual children that were raised in "traditional" (two heterosexual parents) or any type of "non-traditional" family. Those studies are endorsed by professional, scientific organizations.

Bob

Aren't there other opinions about those findings?

Morie

Many of the contesting claims are made by people and organizations that believe that science must conform to their absolute and unquestionable religious beliefs and dogmas. They discount the basic principle of science, that any idea should be open to challenges and could be modified or dismissed if it contradicts testable facts and logic.

Bob

What is wrong with having religious beliefs that are considered absolute, unchallengeable truths, and fitting the scientific observations within those beliefs?

Becky

On its face, that may sound nice, and in the past it was a prevailing attitude. Even today, there are those who accept as true only scientific facts that conform to their religious or ideological beliefs. Personally, I think that ideology should not distort science.

Bob

Why?

Micropsychology

Becky

For several reasons, one of which is the benefits to humankind. For example, if throughout history religious dogmas were successful in limiting scientific research, we would believe today that the earth is the center of the universe. We would not have communication satellites that help connect people around the globe; we would not have weather satellites to warn us from approaching storms. We probably would not have immunization and numerous innovative medical treatments.

Bob

But we would not have atom bombs either...

Becky

People kill, not bombs... While the fruits of science should be used morally, it would be dangerous to treat beliefs as scientific truth. Let me give you an example. After the Russian revolution, the leaders of the Soviet Union were looking for effective ways to change human nature. They wanted everyone to be in line with the communist ideology. They believed that if the current generation were made to follow their dogmas, the next generations would inherit those newly acquired behavioral traits, and obey the new rules. To achieve their goals, they devised and implemented horrendous methods that brought suffering and misery to millions of their fellow citizens.

This created fertile grounds for charlatans and pseudo-scientists, none surpassing Trofim Lysenko. Lysenko, a biologist and agronomist, was looking to improve the weather resistance of agricultural plants and increase their yield. Unlike other scientists that were relying on genetic science to develop new breeds, Lysenko claimed that exposing plants to certain external conditions would make them change their original properties, and that their offsprings would inherit those acquired traits. In the 1920's, this approach matched the ideology of the leaders. Lysenko was promoted and was appointed director of the Institute of Genetics at the Academy of Sciences. Under his reign, research funding for genetics stopped and the science was outlawed. Those who continued to do genetic research on their own were executed, deported to gulags, or imprisoned. His motto was "if you want to obtain a certain result, you will obtain it .... I need only such people as will obtain the results I need." Lysenko substantiated his pseudo-scientific claims by bogus reports and by propaganda. Genetic research was in shambles. Crop-planning policies were based on pseudo-science that conformed to the rulers' ideology. Soon, food shortages and famine were rampant. Millions starved. Ten years after Stalin's death, in 1964, Lysenko was removed from his office, and his theories were denounced.

Morie

Let us go back to our original subject. During the past half century, there has been a surge in the number of "non-traditional" families. If the tens of millions of children who grew up in those "non-traditional" families were more susceptible to homosexuality than children that grew up in traditional families, there should have been a surge in the percentage of homosexuals in the general population. I am not aware of data indicating that there has been such a surge. This is one more reason why I believe that factors outside of the home are major contributors to the development of the sexual orientation of a child.

POSSIBLE GENETIC MECHANISMS

TOKEN-EXCHANGE MECHANISMS

Bob

What brain mechanisms could regulate the development of sexual orientation?

Becky

Let me expand on that question. Behaviors that are controlled by pheromones or hormones are genetic. However, pheromones have not been implicated in human sexual attraction, not even as a co-factor. Hormones, too, are not determining sexual orientation. Are there any other genetic mechanisms that determine sexual orientation in humans? What I mean by such mechanisms is this: genes initiate the production of some physical token in one sex. When members of the opposite sex receive that token, they become sexually attracted to the provider of the token, because their system is genetically designed to act that way. They don't need to learn that token from experiences. The production of the token, its transfer to the partner, and the response that it generates in the partner are all processes originated by genes. This token in humans would play the same role that pheromones play in other species.

Morie

I am not aware of such a token, but I have an intriguing idea of what it could be. The voices of men and women are different in their pitch, and this difference is widely used as a cue for identifying the sex of a speaker, e.g. when we talk to unknown person over the phone. Hair cells that sense sound are located on the basilar membrane in the inner ear. Their sensitivity to the pitch of the sound depends on their location on the membrane.

One end of the membrane is more sensitive to high pitch, and the other end to low pitch (tonotopic response). This sensitivity to pitch is shared also by neurons in a sequence of nuclei that connect the hair cells with the auditory cortex, and by some of the cortical neurons. This entire architecture is regulated by the genes.

In principle, genes could make use of those localized pitch-sensitivities to create neural circuits for selective sexual attraction. In males, the sexual attraction center would be linked to neurons that respond to high-pitch sounds, and in females, the attraction center would be linked to neurons that respond to low-pitch sound. In that way, males will be attracted to females, and females to males. In gays and lesbians, those links would be crossed, and in bisexuals, the attraction center would be connected to both areas of the basilar membrane.

The basilar hair cells are just the first station in sound processing, and they provide the basic information about the pitch of the sound. All the other processing stations and the cortical areas make more specific differentiations between sounds. They could separate human voice from non-human sounds and separate between human voice of high pitch and human voice of a low pitch.

Sol

Oh, I see, so that men will be attracted to women and not to canaries, and women will be attracted to men and not to subwoofers.

Morie

Subwoofers?

Bob

You know, those bass speakers that some dudes have in their cars and play them in full volume.

Morie

Oh, yes.

Becky

The idea that voice plays a part in sexual attraction is compatible with sexual attraction in animals. Many mammals, birds and insects use voice to attract a mate.

Sol

Don't forget the subwoofers…

Claire

However, sexual attraction in humans is not limited to voice; visual stimuli appear to play a greater role than sounds.

Bob

I agree. You may go and ask hundred guys what is the first thing that attracts them to a woman. Very few, if any, would say that it is her voice.

Morie

For the sake of our discussion, let us discuss the attraction of a man to a woman. Similar arguments will hold also for the other types of attraction. While the typical pitch of female voice is always the same, women's appearances change from time to time and from place to place. One way that typical visual features of females could be learned and used as cues for sexual attraction is through their association with the pitch of their voice. When a woman talks, the pitch of her voice identifies her sex. Her visual features, which are coincidental to her voice, can now be learned and serve as visual cues for identifying females. This is as in

Pavlov's conditioning. The female's voice activates the sexual attraction center of the listener, the same as the food pellet activates the salivary glands. Her visual features are analogues to the bell ring. After enough repetitions, typical women's visual features by themselves will activate the attraction center, the same as the bell ring activates the salivary glands.

The pitch of the voice acts like a searchlight at night. It spots targets for sexual attraction. Once in the spotlight, other non-universal features that happen to be associated with the pitch of the voice, such as long women's hair in many cultures, are learned and become new cues for sexual attraction. In later experiences, those newly learned visual cues could serve as spotters of other features. For example, the contours of the female body could become sexual attraction cues through their association with long hair, which has been learned through its association with the high pitch.

Sol

And what about children that are born deaf? How they would learn their sexual attraction cues?

Morie

A voice-based mechanism may be one way of developing sexual attraction, but not the only one. Voice could be central at an early age, because the auditory system can already distinguish between men's and women's voices. At an older age, other mechanisms are probably involved too. Those other mechanisms would regulate the sexual development of deaf children.

Bob

Are you implying that the hearing of heterosexuals is different from that of homosexuals?

Micropsychology

Morie

No. The hair cells in the basilar membranes and the neurons of the auditory track send connections to a variety of brain centers. What I am raising as a possibility is that the connections to one center – the center for sexual attraction – are different in those groups of people. Connections to other brain cells are not affected by the genes that regulate the development of sexual attraction.

Claire

Are there gene-initiated mechanisms for developing sexual attraction, which are based on senses other than hearing?

Morie

In general, there might be other gene-originated mechanisms for learning sexual attraction, similar to the voice-based mechanism. Instead of relying on voice as a token, they could rely on any universal sex-differentiated attribute that has a matching sensor. For example, having or not having facial hair might serve as such a token. Straight males would be attracted to smooth faces and straight females to hairy or stubble faces. Tactile or visual sensors might be innately tuned to such differences, albeit they may not be as simple as the auditory ones. However, it is doubtful that this kind of attraction could be genetic; the sexual attraction tendencies of children of bearded fathers are not different from those of children of smooth-faced, shaving fathers.

Another example might be breasts. If somehow our brains had gene regulated neural circuitry that responds differently to the shapes of men's and women's breasts, or to the differences in the way that they sway, then that circuitry could be innately connected to the sexual attraction center. The entire fundamental sexual attraction process would be gene regulated; males will be attracted to humans with rounded swaying breasts and females to humans with flat solid breasts. If "breast sensors" are not innate, sexual attraction to breasts must be a learned cue. It becomes a cue through association with already established other sexual attraction cues.

Gene-initiated token exchange mechanisms can rely only on universal cues; cues that do not change with time, location, or culture. Cues that are typical to the social environment, such as clothes or hairstyles, could be added to the universal cues by classical conditioning mechanisms.

EMOTION-BASED MECHANISMS

Becky

Can you envisage genetic learning mechanisms of sexual attraction that do not rely on exchange of sensory tokens?

Morie

Yes. By about three years of age, children become aware that there are two groups of people; women and girls is one group and men and boys is the other. This grouping is done based on observations of each child and on guidance given by adults and other children. The distinction between adults and children may be acquired without any external instruction, based on innate detectors that distinguish between small objects and large objects. However, grouping boys with men and girls with women may be the result of explicit teaching.

Children tend to play and socialize with other children of their own sex. This is especially evident right before puberty kicks in, when there is alienation verging on hostility between the two groups. At the same time, most children feel comfortable within their own group. Each of those emotions by itself, and in conjunction with the other, may be the trigger of a learning mechanism of sexual attraction. The emotions of feeling safe within your own group while being threatened by the other are extensions of the feeling SWAP emotion. Now, those emotions guide the formation of the sexual attraction detectors. Features of the adversary group become cues of the activity-plan of sexual attraction. So, for example, a ten years old boy may be terrified by a girl chasing him and threatening to kiss him in front of everyone. As a result of this emotional situation, synapses in his head are formed between the neurons that detected features of girls and the neurons that trigger the activity-plan of sexual attraction. At the same time, any existing connection between detectors of boys' features and triggers of sexual-attraction activity-plans are trimmed off. In the near future, those same girls' cues will cause the boy to be attracted to girls, while boys' cues will have no sexual effects. Similar processes may explain the formation of all the varieties of heterosexual and homosexual attraction in humans.

Claire

Is the tendency of children to form playgroups of their own sex "nature" or "nurture"?

Morie

Boys and girls have both masculine and feminine hormones (androgens and estrogens), but at different proportions. When averaged over large groups, the levels of those hormones correlate with general behavior traits. For example, testosterone correlates with confrontational "masculine" behaviors, while estrogens correlate with "feminine" accommodating behaviors. It is possible that those hormonal levels are regulated by feedback mechanisms; boys engage in activities that keep their testosterone levels high and girls engage in activities that keep those levels low. Therefore, boys will play with other confrontational boys, and girls with other accommodating girls. That may be an underlying genetic design that results in the emergence of the same-sex playgroups.

Becky

Don't you find it a little strange that the system would disconnect from the sexual attraction activity-plans the group that the child is most comfortable with?

Morie

Not at all. This mechanism may show itself also in other related behaviors, such as incest avoidance. There is evidence that this major universal taboo is, at least in part, a natural process that does not require too much coaxing. For example, children in kibbutzim (collective farm communes in Israel) used to be raised in children-homes. They were treated as large groups of siblings – both boys and girls. They spent most of the day, including sleeping at night, together in their children-home. They spent only a few hours a day and weekends with their parents. It was found that as they grew up and left the children-home, those children were not sexually attracted to their group-mates. They were attracted in a normal way to children who grew up in other children-homes in the same kibbutz, and to other people.

Another example is from the other side of the globe. In the past, parents in Taiwan used to pre-arrange marriages for their infant children. In some cases, the bride-to-be was adopted into the husband's family, and the children were raised together. Compared to other families, where the couple were raised separately, those who were raised together had statistically more troubled family lives, with weaker sexual attraction between the couple, that resulted in less offsprings.

Sol

In our discussion of voice-based learning mechanisms, we relied on the fact that different parts of the basilar membrane respond differently to different pitches. This translates into anatomical regions that respond differently to males and females. That difference could be used by the genes to construct mechanisms that would train the sexual attraction center to respond to a specific sex. If we want to extend that idea so that emotions replace voice, some parts of the brain would need to act differently when we experience different emotions, in analogy to the basilar membrane. That would enable the genes to link the sexual attraction center differently to males and females, according to the different emotions that they invoke. Do different emotions activate different parts of the brain?

Morie

In general, several brain regions are implicated in generating and processing emotions, including the limbic system, the amygdala and the orbitofrontal and anterior cingulate gyri. At birth, the limbic system is the most developed of those regions, and it is responsible for executing the bodily expressions of basic emotions such as enjoyment, fear and anger. All those brain regions continue to develop, generally from the lower limbic system outwards to the cortex. As they develop, the brain becomes capable of expressing emotions that are more complex, being aware of them, and associating them with external events.

There is some asymmetry between the right and the left hemispheres of the brain in processing emotions. The left hemisphere is more implicated in good feelings and the right with ones that are more negative. For example, people who suffer damage to the left frontal lobe become more depressed, while similar damage to the right lobe often results in unwarranted cheerfulness. The roles of the two hemispheres may switch back and forth, as a child grows up, and it varies from one individual to another.

All those emotion regions are interconnected among themselves and with other brain regions. It is known that injuries to certain brain regions result in changes of behavior that include changes of emotions, such as from aggressiveness to docility. The bottom line is that a comprehensive mapping of emotions onto brain regions or onto types of cells is not yet known, other than in general outlines.

However, since the development of emotion loci is gene-controlled, it is feasible that certain emotions could be used as spotters for external cues that have to be incorporated into the sexual attraction centers. For example, the emotion felt by a boy who is being taunted by a group of girls may cause the incorporation of the general characteristics of the girls into the boy's developing sexual attraction program, and thus shape his sexual orientation. That emotion could be a combination of the fundamental emotions of disappointment and anger.

Claire

The voice-based mechanism can function at an early age, and it does not require that children have any knowledge or awareness about the sexes, except that there are two groups of adults, one speaking with high pitch, the other with low pitch. Voice-based and similar mechanisms can build up the sexual attraction of children even before they are aware of their own sex. Those buildups are done autonomously by the brain, and they do not require cognitive intervention of the child, or being taught by adults.

Micropsychology

Morie

That's right. These are precursors of sexual attraction, which will become completely functional at around puberty.

Claire

And what happens when the child reaches puberty?

Morie

The actual sexual attraction of an individual will be determined by the combined effect of all the autonomic learning mechanisms that took place from right after birth to puberty. As we said, at around three years of age children become aware that there are two groups of children; boys and girls, and each child belongs to one of the groups. The understanding of the full meaning of the concept "sex" has not yet reached its maturity in the growing child. Sexual attraction could be developed autonomously by voice-based and equivalent mechanisms, as prescribed by the genes and influenced by individual experiences. Those mechanisms recruit cues that are taken from the adults' morphology.

When children reach puberty, their sexual attraction activity-plans mature. What used to be boys and girls are now men and women. The sexual attraction activity-plans in their brains become operational on one hand, and on the other hand, the young adolescents now send out the cues to trigger those programs. They feel sexually attracted to their peers and to other adults.

Many people report that when they had reached puberty they suddenly found themselves sexually attracted to the opposite sex, without noticing how that feeling developed. Others report that they found themselves attracted to their own sex, while others felt ambiguous. Many people also report that they felt vague, unexplained attraction throughout their entire growing up, and that the feeling crystallized into sexual attraction as they reached puberty. Most report that their sexual orientation was already fixated by the time they reached puberty.

Claire

How may fixation take place?

Morie

I believe that this fixation is done through imprinting mechanisms. Each process of learning sexual attraction and arousal has a certain time window during which it has to occur and at its end, it is fixated. Once it is fixated, it cannot be deleted.

Claire

If not completely deleted, could it be modified?

Morie

It could be expanded, but not deleted. For example, once a man is heterosexual, the features that attract him to a woman may be modified, but usually not his attraction to women.

Bob

Have there not been cases in which people changed their sexual orientation?

Morie

There have been. Both men and women that considered themselves heterosexual and had traditional families and children became later in their lives homosexual. However, it would be difficult to determine that those people were not bisexual or borderline bisexual to start with. The same holds for homosexuals that later in their life became heterosexual.

Micropsychology

Becky

You have described general processes that underlie the development of sexual attraction and sexual arousal. You have said that the micropsychology of those processes is the same as that of all other non-sexual mental processes. What are the biological building blocks of all those processes?

Morie

That is going to be our next main topic.

CHAPTER TWO

NEURONS AND INFORMATION REPRESENTATIONS

Morie

Today we'll deal with the basic mechanisms that underlie the processes that we discussed earlier. Those mechanisms are not restricted to the sexual arena; they underlie many other conscious and unconscious brain processes.

THE NEURON

There are about hundred billion neurons in the brain. Like trees in a forest, neurons have various shapes and metabolisms, but they all share some basic features.

A neuron has three main parts: dendrites, a soma, and an axon (Figure 1). The dendrites resemble branches of a tree. They join the soma, which, like a tree's stem, is the central, bulky part of the neuron. Similar to a root that leaves the stem, a single axon leaves the soma, and divides into a hierarchy of axonic branches.

Apart from this visual resemblance, a neuron is nothing like a tree. A neuron is an electric relay. Pulses of electricity travel in the dendrites towards the soma. (Arrows in Figure 1 indicate the directions of the pulses.) The soma collects those pulses and adds them up, and if they pass a certain threshold, the soma sends its own electric pulse down the axon. This pulse spreads to all the axonic branches. When the soma relays such a pulse, we say that the neuron fires. Otherwise, the neuron is at the quiet state.

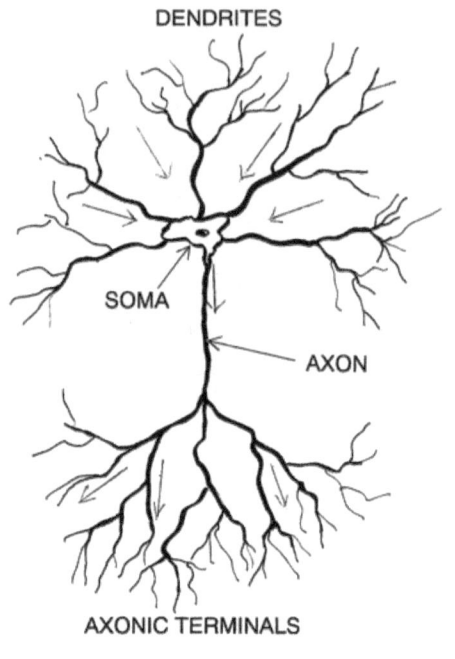

Figure1 The Neuron

Sol

What initiates the electric pulses in the dendrites?

Morie

It depends on the type of the neuron. Sensory neurons have sensors at the tips of their dendrites. The sensors convert external stimuli to electric pulses. Each type of stimulus has its specialized sensor. For example, when pressure is applied to the skin, it activates pressure sensors, which send electric pulses down their dendrites. When light enters the eye, light sensors in the retina send electric pulses through their dendrites to their somas.

Sensory neurons are found in the peripheral nervous system. The vast majority of the brain's neurons, though, are interneurons. Interneurons do not have sensors. Instead, axons of other neurons are connected to their dendrites. Firing of those axons induces electric pulses in the dendrites of the interneurons.

The connection between two neurons, a source and a target, is called a synapse. A synapse is a narrow gap between an axonic terminal of the source and its target, usually at the surface of a dendrite.

As the source fires, the electric pulse reaches its axonic terminals and causes vesicles in them to release chemical compounds, called neurotransmitters. The neurotransmitters cross the synaptic gap to the target neuron. If enough of them have crossed, they would generate a new electric pulse in the target. The new pulse propagates to the soma, where it is collected.

The intensity of a pulse as it starts is propagation in the target neuron depends on the efficacy of the synapse by which it was relayed. Different synapses may release different amounts of neurotransmitter. This property is referred to as the *synaptic weight*. The higher the synaptic weight, the higher is the efficacy of the synapse.

Several dozens of neurotransmitters have been identified in the brain. Usually, every neuron releases a certain neurotransmitter. However, a neuron may receive various neurotransmitters, and integrate their contributions.

Some neurotransmitters reduce the intensity of the new electric pulse, or even prevent its formation altogether. These are called inhibitory, whereas the others are called excitatory.

Overall, the brain consists of networks of neurons that send electric pulses to each other.

Sol

What is the purpose of all those pulses?

Morie

There are two main purposes. First, these pulses represent information, and by manipulating them, the brain processes its information. Second, a group of neurons, called motor neurons, have axons that connect to muscle units and to glands. Firing in the motor neurons activates their muscle units and the glands. So, the electric pulses that the motor neurons send trigger all our physical activities.

REPRESENTATIONS

Claire

What did you mean when you said that the electric pulses represent information?

Morie

Everything that we sense, feel or do, consciously or unconsciously, is represented by groups of firing neurons. In this respect, the brain is like a huge control board of a plant. At any time, different light bulbs are lighted. Those lighted patterns represent what processes are taking place in the plant.

Let me elaborate on that. Consider an event in which someone touches you on the shoulder and you turn your head in that direction. First, pressure sensing neurons in the touched area fire. Each of them is a representation of its touched spot. Their pattern, as a whole, represents the entire touched area.

That information passes to intermediary neurons that evaluate the situation. Different groups of those firing neurons represent concepts that the brain uses in the evaluation process. For example, a group of firing neurons represents the 'novelty' of the touching. Another group represents the concept 'need to find more information about the touched area'. Eventually, pulses reach the motor neurons that fire and elicit a response. Each firing motor neuron represents its activated muscle unit, and their entire pattern represents the turning of the head.

Bob

That raises an intriguing possibility. If we could monitor every neuron in the brain of a person and identify the firing neurons, we would know what concepts the brain is handling. That means that we would know what the brain is sensing, feeling, thinking, planning and so on. Would it be possible to build a mind-reading-machine, by monitoring the firing of neurons?

Morie

In principle, you may be right. In practice, I would not bet on it. Here is why: First, there are billions of neurons in the brain. I do not think that it will be possible to monitor a significant number of them anytime soon. Remember, the monitoring has to be non-invasive and non-destructive.

Second, the brain uses a huge number of concepts. However, we only know how very few of them are mapped onto patterns of firing neurons. The mapping of most concepts is still terra incognita.

Bob

What about those colorful PET-scans and MRI's of the brain, which we see on TV? Don't they show how the brain works in various situations?

Morie

Those advanced techniques are very useful, but their resolution is limited. They can show how regions of the brain are involved or not involved in certain general activities, but they cannot zoom-in on single neurons, or even on small groups of neurons.

Bob

Is there any evidence to support the idea that firing neurons represent concepts?

Becky

Let me answer the question; I have learned about it. The electric state of a neuron can be measured by an electrode (a thin needle) inserted into it or positioned next to it. It is possible to detect when a neuron is collecting electricity and when it fires. By inserting electrodes into neurons at various brain regions and stimulating various body parts, correlations were found between firing neurons at various brain regions and those body parts. The entire surface of the body was divided into small segments – pixels – and neurons that fire in response to specific pixels were identified and mapped. Each such neuron is said to represent its body pixel. Those neurons are organized in a brain area called the sensory part of the sensorimotor cortex.

Similar correlations have been found between sensory neurons of other senses such as vision, hearing and smell and neurons at corresponding brain areas.

It was found that certain brain neurons fire in response to specific patterns of activated sensory neurons. For example, certain neurons in the visual cortex fire in response to illuminated lines that cross the field of view. Some neurons fire selectively when lines pass at a certain speed and orientation. Other neurons fire in response to lines of certain width.

Experiments that were done on monkeys support the idea that even complex concepts are represented by firing neurons.

Monkeys were shown pictures of other monkeys. Some pictures showed the head of a monkey staring straight ahead, and other pictures showed a head of a monkey in profile, looking to the side. As a picture was shown, electrodes scanned the visual area of the monkey's brain. It was found that certain neurons fired selectively in response to the pictures of the staring monkey, and other neurons fired selectively in response to pictures of a monkey looking to the side. That suggests that certain neurons represent the concept 'another monkey is staring at me' and other neurons represent the concept 'another monkey is looking to the other direction'. For a monkey, these are two distinct, significant concepts that may call for different responses.

In humans, a certain part of the lower temporal lobe of the brain has also been implicated in facial recognition. Lesions to this region impair one's ability to recognize even familiar faces.

Correlations have been found between the firing of certain neurons and activation of muscle units. Those neurons represent the activity of those muscle units.

Firings of neurons in other parts of the brain were correlated with a variety of activities and emotions such as anticipation, fear, anger, and pleasure.

It seems that there is ample evidence to support the idea that firing neurons represent concepts. However, it is not known what exact patterns of firing neurons correspond to most of the concepts that the brain uses. The code that the brain uses to represent concepts by its firing neurons is still not deciphered.

BRAINS AND COMPUTERS

Sol

The way that the brain represents information reminds me how information is encoded in computers; by zeros and ones. A firing neuron is like a one, and a quite neuron is like a zero. Does the similarity go beyond that?

Morie

There are some other similarities but there are basic differences. For example, the fundamental manipulation of information in computers is done with bytes, which are relatively short sequences of zeros and ones, while in the brain millions of firing cells could be processed at once. Also, unlike computers, which store information in random access memory (RAM), the brain's memory is associative memory, and information is stored in it as content addressable.

Bob

What is the difference between RAM and associative memory?

Sol

Let me give you an example. Assume that you go on a blind date. Meeting your date is like retrieving information; you have to know how to get to it. One way would be to use the physical address of your date: street name, house number, and floor and apartment number, if applicable. This is the basic idea of RAM. In order to retrieve information you have to know its address. Like houses in a city, cells in the computer's memory have addresses. Information is placed in those cells arbitrarily, and is retrieved by its cell address. Like an apartment that can house different people at different times, a cell of a RAM can store different information at different times.

Another instruction for meeting your date may sound like: I'll be at the lobby of Concert Hall, not far from the entrance. I have brown hair and I'll be wearing a brown scarf with white stripes. The first part of the instruction is an address, albeit not a specific one. The second part, the hair and the scarf, are parts of the contents of the retrieval target. These are cues that you are familiar with and with which you can identify your retrieval target. This part of the instructions is content addressable. This is also an example of associative memory. An item (your date) is retrieved based on its association with a given item or items (hair, scarf).

Morie

The brain uses similar strategies to store and retrieve information. Various parts of the brain store information of specific types. There are areas for visual information, areas for auditory information, areas for information of combination of senses, and so on. Within those general areas, retrieval is by the content of the retrieved, or put in different words, by associations that exist between the retriever and the retrieved.

Bob

We have seen examples how concepts could be encoded and retrieved in a neural network. Do genes control the development of such complex structures?

LEARNING IN THE BRAIN

Morie

There is no difference between the brain and any other organ of the body; their development is controlled by genes. Genes control the development of the brain in the fetus and after birth. They control the creation of individual neurons and the formation of the networks. Newborns already have almost all the neurons that they will have as adults. However, connections between neurons continue to develop throughout life.

Sol

By controlling the structure of the neural networks, genes determine how signals would flow in our brain and the resulting neural firing patterns. Since those patterns represent concepts, does it mean that the genes determine the concepts in our brains, even before we are born?

Morie

In addition to building the neural networks of the brain, the genes provide it with mechanisms to what is called *"changing synaptic weights"*. This includes cutting off connections between neurons, adding new connections, and changing the efficacy of synapses. Changes in synaptic weights depend on external information that enters the brain. Those changes can modify the patterns of signal flow in the brain. That, in turn, could cause quiet neurons to fire, or firing neurons to become quiet. Thus, representations of new concepts are formed by the modification of synaptic weights.

FROM NEURONS TO NODES

Bob

You mentioned that the mapping of concepts in the brain is terra incognita. If we do not know even the code that the brain uses to record its concepts, how can we proceed and talk about the underlying biological correlates of human behavior?

Morie

Not knowing the mapping and the code limits the detail-level of our descriptions, but it still allows us to study and understand basic brain mechanisms. Instead of describing the activity of each individual neuron, we will talk about the activity of groups of neurons. Each such group will be represented by a *node*. A node represents a pattern of neurons that underlie a concept or perform a specific activity. We may not know the details of the connections between the group of neurons that represent the concept or how they perform their specific activity, but we know that somehow they do it. The node, and not the individual neurons, is the smallest element in our model.

For example, the entire brain circuitry that can recognize the flavor of a banana would be represented by a node. By dealing with nodes, we bypass our lack of knowledge of such circuits, and we rely only on their observed functionality.

The nodes represent the concepts that the brain uses. They represent our physical and mental worlds. The connections between nodes encode relationships between the concepts that they represent.

In addition to representing concepts, nodes can do things that are done in the brain by specialized circuits of neurons, such as controll walking. Again, we may not know how exactly the brain does those things, but in the resolution level of our model we do not deal with those details.

An important group of nodes is responsible for processing information. They organize the recording of new information and the retrieval of existing information from the memory.

A node has input channels (dendrites), output channels (axonic branches), and a body (soma) that integrates incoming signals and decides if the node should fire. A firing node means that the concept that it represents is active.

Unlike neural synapses that transmit signals in one direction, signals between two nodes can propagate in both directions.

In real neurons, neural synapses connect an axon of one neuron to a dendrite of another. In nodes, in addition to representing neural-like synapses, nodal synapses represent any interaction between two nodes. Real neurons may interact through a third party, which is not a neuron. For example, a firing neuron may activate a gland, which releases hormones, which affect the firing of other neurons. In the nodal network, the two nodes would be connected directly, and the effects of the hormones would be represented by the synapse that connects the two nodes. The details of the communication mechanisms are not addressed by the nodal network.

Describing the activities of the brain by nodes is like describing a war by the activities of battalions, instead of by the activities of all the individual soldiers. Both battalions and individual soldiers fire their weapons. However, battalions include also soldiers that do not fire – their role is to support the soldiers that fire. Generally, we can follow a war by following the activities of battalions.

Bob

It makes a lot of sense. In life, we do it all the time. We use things even if we do not know the details of their structure and operation. Cases in point: watches, cell phones, TV and more.

THE BASIC TYPES OF CONCEPTS AND THEIR ORGANIZATION

Morie

Imagine opening your eyes in a winter morning to the serene sight of snowflakes falling quietly. Millions of white dots fill your field of view, stimulating millions of neurons in your eyes and brain. In no time, the brain processes all those millions stimuli and comes up with one concept that summarizes the event: 'It is snowing'.

The brain deployed two main processes: First, it determined that all those stimuli belong together and could be treated as one item. This item may be called 'many falling snowflakes'.

Next, the brain gave a meaning to that item. It determined that it is similar to other events in which snow was falling, and identified it as an exemplar of the class 'it is snowing'.

These are staple processes that the brain uses when it handles information. It lumps together many stimuli into one item, and then assigns that item to a class that contains similar items.

These processes depend on two relationships between concepts that the brain handles. The first is the relationship between parts and their item; the parts make up the item. The second is the relationship between exemplars and their class; similar items are exemplars of the same class.

The concepts are represented by nodes. The relationships between concepts are encoded by the synaptic weights between their nodes.

Sol

Let us reconsider a falling snowflake. We said that it is a part of the item 'snow is falling'. However, 'snowflake' is also an item that has its own parts; the tiny icicles that make it.

Morie

That is correct. The brain organizes items and their parts in hierarchies. An item, such as 'snowflake', could be a part of other items, such as 'snow is falling', and have other items as its parts, such as 'tiny icicles'.

You mentioned that a snowflake has tiny icicles as its parts. The brain may consider also general properties of an item as its parts. For example, the brain identifies a falling snowflake by its size, color, typical drifting in the air, and its similarity to the multitude of other falling snowflakes. All these properties are considered by the brain as parts of 'snowflake'.

Sol

What about classes and exemplars; are they also organized in hierarchies?

Morie

Yes, they are. 'It is snowing' serves as a class that one of its exemplars in the current snow fall. At the same time, 'is snowing' is also an exemplar of the class 'weather conditions', which includes other exemplars such as 'it is raining' and 'it is shining'.

Sol

I have another comment. In addition to all those relationships, every snowflake is an exemplar of the class 'snowflake'.

Morie

You are correct again. A concept can be an item, a part, a class and an exemplar. It all depends on its relationships with other concepts. The brain organizes its concepts in items-parts and classes-exemplars hierarchies. A concept may belong to several hierarchies.

The situation may be summarized as "The brain represents information as inter-connected items-parts and classes-exemplars hierarchies".

Bob

I must admit, I understand better things that I can see. Those inter-connected hierarchies are a little too abstract for me. Is there a way to visualize them?

Morie

We can describe the information structures of the brain by diagrams. Concepts are represented by boxes. Relationships between concepts by arrows: A pointed arrow denotes the relationship between an item and its part, and a circular arrow denotes the relationship between an exemplar and its class. As an example, Figure 2 illustrates portions of the information structure that includes the concept 'snowflake'.

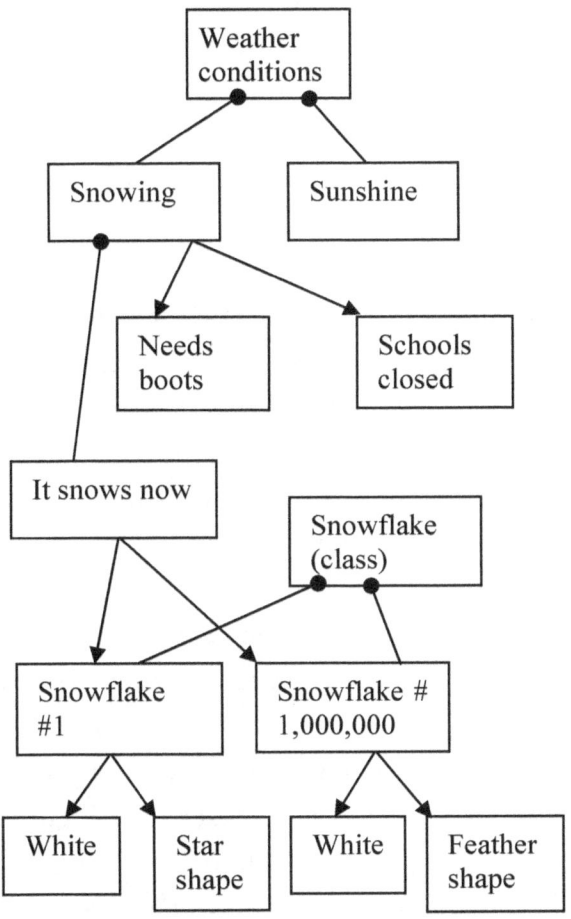

Figure 2 Inter-connected information hierarchies.
Item part   ⟶
Exemplar class   ————●

Claire

What are the benefits of organizing information in such hierarchies?

Morie

Tens of millions of sensory cells, relentlessly, bombard the brain with information. The brain must reduce that onslaught to a manageable number of information entities. That is why it assembles parts to items and organizes them in hierarchies. Now, the brain has to deal with a smaller number of information entities – the items that represent the current situation.

In order to benefit from its past, the brain needs to have quick access to records of similar experiences. The brain accomplishes it through its class-exemplar hierarchies. A class includes three lists: First, a list of similar items, which are the exemplars of that class. Second, a list of conditions, or cues, that would qualify an item to become an exemplar of that class. Third, a list of properties that are shared by all the exemplars of that class. Once a new item is assigned as an exemplar to a class (based on its own properties, which served as qualifying cues), it inherits the properties that are shared by the exemplars of that class.

For example, based on the falling snowflakes, the current condition was identified as 'it snows now'; many parts were assembled into one item, which represents them. Based on its cues (many falling snowflakes), the item 'it snows now' became an exemplar of the class 'snowing'. It inherited the properties of this class: 'needs boots' and 'schools are closed' (see Figure 2).

The organization of information into those hierarchies provides the brain with an efficient way of accessing and handling information.

DEVELOPMENT OF INFORMATION STRUCTURES

FROM PARTS TO ITEMS, EXEMPLARS AND CLASSES

Morie

The brain develops its information structures all the time by modifying synaptic weights. It assembles parts into items and assigns items as exemplars to classes. It also defines new classes and establishes their relationships with existing ones.

In addition, the brain singles out typical feature combinations that could serve as cues, by which it identifies and classifies new items.

For example, one of the first things that babies learn is to recognize their mother. Innate cue-detectors detect discontinuities and movements in the visual field. They detect the outline of her hair, the darker spots of the nostrils, the eyes, the moving lips that surround the darker opening of the mouth, and so on. Nodes that represent elements of those features become part nodes, and they recruit item nodes to represent them. There are nodes that represent the eyes and their parts, the lips and their parts and so on.

An item node that represents the concept 'mother' is then recruited. Its parts are all those newly formed feature combinations, which serve also as cues to the concept 'mother'. Whenever the mother appears, those cue nodes are activated, they activate the node 'mother', and the baby is aware of the presence of the mother.

With time, more stimuli appear in conjunction with the old ones. They include typical sounds such as the mother's voice, her steps, her softness, and so on. The brain adds new cues to the concept 'mother'.

Then, other people enter the world of the baby. Their presence triggers some of the mother's cues, but also some new cues. The brain represents those people by new item nodes, each of which has as parts the cue combination that characterizes that person.

Some cues turn to be parts of both 'mother' and nodes that represent other people. That prompts the brain to define a new class node 'person', whose exemplars are 'mother' and nodes that represent the other people. The cues of the class 'person' are the cues that are common to all the exemplars. These cues continue to serve as parts of 'mother' and the other people, but they are not anymore specific cues to those items. Cues that activate the node 'person' would not activate the node 'mother' by themselves. Some cues that are specific to 'mother' would be needed too.

The brain keeps adding exemplars to the class node 'person', according to the experiences of the baby. After a while, cues that are common to some of the exemplars recruit their own class nodes. Classes such as 'big person', 'small person', 'person with prickly face' and 'person with smooth face' are formed. They become exemplars, or subclasses, of the class 'person'.

The information structure that includes all the various types of human beings is learned at a very early age. It takes part in forming many other concepts that involve the interactions of the individual with other human beings.

Thus, it all starts with a set of innate cue detectors that detect feature combinations that occur in the world. When such cues are detected, they define new items. Similar items then define classes. Similar classes define their supper classes, with their own cues and exemplars, and on and on.

The innate cue-detectors are determined by the genes, but all the other cues, items and classes result from the interactions between the innate cues and the experiences of the individual.

CHANGING SYNAPRIC WEIGHTS

Sol

How the brain determines which synapses to modify, in order to build its information structures?

Morie

The mechanisms of modifying synaptic weights are genetic. Only synapses that relay pulses from a firing node can be modified. The nodes that receive the pulses may fire or be quiet.

For example, when a part node recruits an item node, the part node is firing and the future item node is quiet.

In Pavlov's conditioning, a synapse is modified between two firing nodes. Both the node that represents the sound of the bell and the node that causes salivation are firing, and the synapse between them is modified.

When the synaptic weight is increased, the association level between the nodes increases. When the weight is decreased, the association decreases.

Sol

Is the process of synaptic weight modification purely local, meaning that it depends only on the firing status of the two involved nodes?

Morie

In some cases, it is purely local and in some cases, other brain areas are involved.

For example, after birth, many motor neurons activate single muscle units. After a while, the brain trims ties from motor neurons that have fired to target muscle units that remained quiet – "use it or lose it". This seems to be a purely local process.

On the other hand, in Pavlovian conditioning other brain areas are involved in changing the synapse between the firing nodes of the bell's sound and the salivation gland (Figure 3). The weight increases if food was present, and decreases if the dog was expecting food and food was not provided. The first is a positive experience and the second is a negative. The decision to increase or decrease the synaptic weight is reached in other brain areas, and they broadcast it. The firing neurons pick up the message, and modify their synapses accordingly.

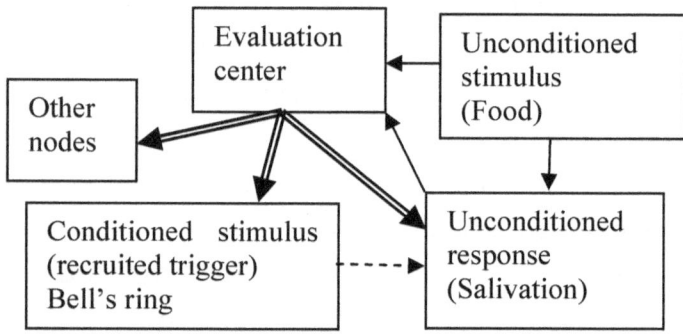

Figure 3 Communication pathways between nodes in Pavlovian conditioning
Pulses from a firing node ⟶
Broadcasted instructions (increase or decrease weight) ⟹
Modified synapse - - - -▶

In general, brain units that evaluate an event broadcast instructions that reach many nodes. However, the instructions may only affect synapses of firing nodes. Such synaptic weights may increase or decrease, depending on whether the experience was positive or negative.

The instructions can also affect the level of the modification. The changes may be small or large, according to the urgency of the modification.

Sol

Is there evidence that synaptic modifications take place when the brain records new information?

Becky

The idea that the synaptic weights of a neuron may change as a result of its firing has been demonstrated in what is called "Long Term Potentiation" (LTP) and "Long Term Depression" (LTD). In both cases, measurements were taken from two neurons connected by synapses; the pre-synaptic and post-synaptic neurons. The pre-synaptic neuron was stimulated by an electrode, and the induced electricity at the post-synaptic neuron was recorded by another electrode. It was found that in LTP the effectiveness of the synaptic transmission increased with repeated stimulation of the pre-synaptic neuron, while in LTD it decreased. Those changes lasted for several hours after the pre-synaptic stimulation ceased. Such experiments were done in vitro in slices of brain tissue from the hippocampus.

It was found that the speed at which synaptic weight modifications happen could be manipulated by introducing various neurotransmitters to the vicinity of the synapses. That opens the door to the possibility, in principle, of improving learning by drugs. If it were possible to administer neurotransmitters to spots where synapses are being modified in a learning process, the pace of the learning could be controlled. If you took a math-learning-pill, you would absorb an idea the first time you read about it. No repetitions would be required.... Unfortunately, this is not available yet.

In several situations, learning has been associated with observed changes in the connectivity of the underlying neural networks. For example, at about eight months of age, synaptic density in the cortical visual area V1 reaches its maximum. Then, from around the age of two years until late childhood, synapses are eliminated. In this synaptic pruning, about forty percent of the synapses are eliminated. Those changes coincide with the improvement of visual skills such as acuity and binocularity. That is one of many indications that synaptic modifications are implicated in changes of functionality and behavior.

An important feature of many of those experiments is that certain modifications of the network's connectivity can occur only within a specific time-window. If the animal was deprived of stimuli during that critical time-window, it could not acquire the corresponding skill afterwards, when the stimuli were re-introduced. If, on the other hand, the deprivation occurred only after the critical time-window, the animal did not lose the acquired skills.

EXAMPLES OF SEXUAL INFORMATION STRUCTURES: RAMON AND JULIE

Sol

A main premise of the model is that all information is learned and stored in the brain by the same mechanisms and according to the same rules and strategies. Could you please give examples how information related to sexual arousal and attraction is learned and stored?

Morie

I am not aware of any scientific experiments that document the development of sexual behavior at the neural level. In particular, which patterns of firing neurons are responsible for certain behaviors, and how those patterns evolve.

Bob

So, for illustration purposes, why not talk about hypothetical persons, and describe how such processes might take place in their brain?

Morie

OK. Let's start with cues that drive sexual arousal. Almost every part of the human body, including chest, arms, legs, buttocks, hair, eyes, ears, fingers and toes has been reported as a sexual arousal cue by some men and women. A large number of inanimate objects such as boots, high heel shoes, lingerie, underwear, leather clothes and many more also act as turn-on. The common to all those cues is that they are acquired through individual experiences. Let's call our hypothetical guy Ramon, and say that he is aroused by women's legs. We want to understand how Ramon learned to be sexually aroused by women's legs. The following is a suggestion of how that could have been evolved in his case.

By all accounts, Ramon was an average child. In the presence of intimidating strangers, he would seek refuge at his mother, clinging to her legs as a toddler, or staying next to her when older. Little Ramon's emotions in those circumstances may be described as 'feeling protected while vulnerable'. Earlier we dubbed this feeling as SWAP, for Safe With Another Person. SWAP is a combination of basic universal feelings, and the circuitry that supports them is probably innate. Learning the association between the external stimuli of touching his mother's legs and the internal feelings of SWAP is the first step in the acquisition of touching legs as a cue to sexual arousal.

101

Conditioning is a central mechanism in this acquisition. SWAP is the unconditioned stimulus that is activating the unconditioned response of sexual arousal. The recruited arousal-trigger is 'touching legs'.

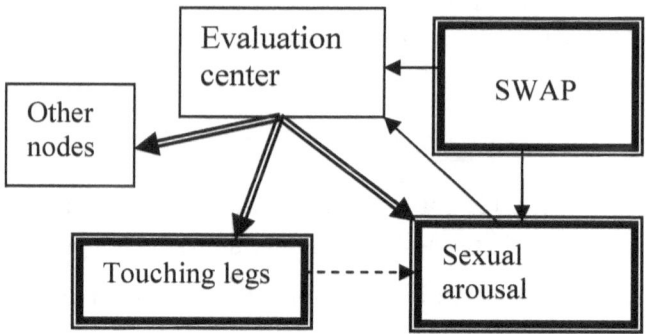

Figure 4 Recruiting sexual arousal cues through conditioning (compare Figure 3). Highlighted boxes indicate firing nodes.

Figure 4 illustrates the circuitry in Ramon's brain. It is the same circuitry of Pavlov's experiment (Figure 3) except for the reassignment of different contents to some of the boxes. The learning mechanism is activated when Ramon feels SWAP. This learning mechanism guides the creation of a connection from 'touching legs', which happens to be Ramon's coincidental activity at that time, to the node of 'sexual arousal'. As a result, 'touching legs' becomes a cue of 'sexual arousal'. The nodes that were firing during that learning process are marked by double frames in figure 4.

The item-node 'touching legs' represents Ramon's mother's legs. At this age, little Ramon's sexual arousal machinery is not functional yet. Therefore, he is not becoming sexually aroused when clinging to his mother's legs.

Ramon would have other experiences in which legs of adults were associated with him feeling SWAP. For example, in events that his father or a nursery school teacher protected him. All those associations were recorded too by the system. The result was that the item node 'touching legs' became a class node. From now on, touching any legs would send signals to the node of sexual arousal.

So far, 'touching legs' is not gender-specific. Some gender-specific cues have to be added to it, so that Ramon will be aroused only by women. It is not known what gender-specific cues are used by the brain to identify the gender of the arousal source (women in Ramon's case).

It has been suggested earlier that the pitch of the human voice may serve as such a cue, because both its generation and its detection are done by body parts whose developments are regulated genetically. Baby boys would be genetically pre-disposed to become sexually attracted to women's voice and baby girls to man's voice.

However, other cues may be identifying women in conjunction with feeling SWAP. The detectors of those cues may be innate, or they may be a generalization of genetic ones. Whatever the origin of the detectors happens to be, they would narrow Ramon's sexual arousal trigger from 'touching legs' to 'touching women's legs'.

Bob

Since Ramon is our average person, I would assume that other cues, in addition to women's legs that were associated with him feeling SWAP, would arouse him. How those other cues may have been formed in his brain?

Morie

One main mechanism is generalization. In generalization, a class node takes on a role that is originally played by its cues or its exemplars. In Ramon's case, the class node 'woman' becomes a trigger of 'sexual arousal', a role that was originally played by its cue 'women's voice'. According to the fundamental mechanism of synaptic weight modifications, a synaptic weight between 'woman' and 'activator of sexual arousal' is strengthened. This happens because both of them have fired in response to the firing of 'women's voice' (Figure 5).

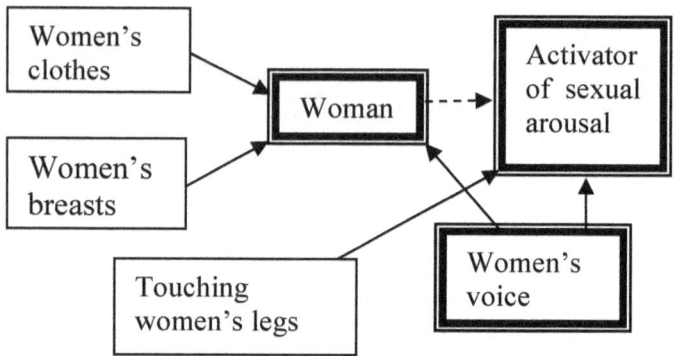

Figure 5 Generalization. The class node 'Woman' becomes a cue that activates sexual arousal, as a generalization of the cue 'women's voice'.

Other cue nodes of the class 'woman', such as 'women's clothes', activate 'woman', which triggers 'activator of sexual arousal'. Through this process of chain activation, they too become triggers of 'activator of sexual arousal'.

Bob

As an average person, there would be women that will not attract or arouse Ramon; for example his mother, sister, very old women, very young girls and most likely very ugly women. How are those eliminated from his list?

Morie

Technically, it is not difficult for the network to prevent certain exemplars of the class 'woman' from activating the sexual arousal and sexual attraction nodes. This could be accomplished by inhibitory synapses. For example, strong inhibitory synapses from the node that represents Ramon's mother to the node that activates sexual arousal would prevent the latter from firing in response to 'Ramon's mother', even if it gets excitatory signals from the node 'woman'. The rest of the brain does not have to be aware of those conflicting instructions ('activate arousal' that came from 'woman' and 'inhibit arousal' that came from 'Ramon's mother').

However, that is not the only possible network architecture. It is possible that a control circuitry checks the potential risks and benefits of sexual arousal, before the person becomes sexually aroused. After weighing the risks and benefits, the circuitry may decide to execute or to reject the arousal. The rejected option may linger in the system, trying to get an approval – a situation that may create mental distress.

Most animals avoid incest naturally, and not because of a social taboo. It is plausible that detectors of incestuous situations inhibit the nodes that activate sexual arousal. Those inhibitory signals may be triggered by pheromones of kin identification.

In humans, learned kin features may be used for incest avoidance. For example, features of a person who is continuously present around the growing child may become cues that that person is a kin. Those cues would suppress sexual arousal.

Other social taboos, as well as cognitive decisions about sexual behavior, are processed by circuitries that weigh the risks and benefits of those activities. Ramon's actual behavior would depend on his personal experiences and the wiring of his brain's circuits that process this information.

Bob

Let's assume that Ramon has a fetish. Even mainstream people that have had an apparently normal childhood may have one. How would this be expressed in the nodal network that controls his sexual behavior?

Morie

Oh, yes. "Fetish—bizarre to the outsider, bazaar to the insider". Assume that Ramon's fetish is to lick women's boots, and that this fetish is in addition to all his other sexual habits. Let's call his partner Julie.

So far, we described (Figure 4) how the innate node SWAP served as a detector that activated the learning process, which acquired 'touching women's legs' as a sexual arousal cue.

The node SWAP can serve also in another capacity – as a class node that activates sexual arousal. Its exemplars would be all the various life experiences that evoked the feeling SWAP in Ramon. For example; little Ramon complies with his mother's demands not to run wild in the store. He stays next to her, aware of her authority and knowing that he is safe as long as he behaves. This entire situation becomes an exemplar of 'SWAP'. Other similar experiences too become exemplars of that class. Eventually, the class SWAP develops new cues. One of the cues is 'obeying the authority results in feeling protected'.

In all cultures, bowing and kneeling are symbols of obedience. Ramon picks it up, and it becomes an indirect cue to SWAP. When he kneels to Julie, it expresses her authority over him. His licking of the boot is a specific cue that activates the concept 'total obedience'. All these actions trigger in his brain the node 'SWAP', which triggers sexual arousal.

Claire

Are all Ramon's sexual behaviors affected, directly or indirectly, by the node 'SWAP'?

Morie

No. So far we have discussed behaviors that could be acquired at any age; both before and after puberty. They could be acquired based on experiences that do not involve mature sexual activity. After puberty, a new player joins the game. As Ramon engages in sexual activity and enjoys it, that enjoyment triggers the learning of new sexual arousal cues. This is according to the general rule that emotions trigger and facilitate learning. For example, after having one gratifying sexual experience, just being with a woman in bed may become a cue for arousal. Direct synapses may be formed from the concept 'I am in bed with a woman' and the activators of 'sexual arousal'. The feeling of 'SWAP', or its derivatives, will not be mandatory for Ramon to be aroused.

Claire

We have seen that several processes and mechanisms are involved in learning and organizing sexual cues. Could it be possible to summarize them in an illustration?

Morie

The illustration would look "busy", but it would be worthwhile to study it carefully.

Explanation of Figure 6

The figure shows the connections between some of the nodes in Ramon's adult brain that have been involved in his sexual arousal and attraction.

Figure 6 (see text)

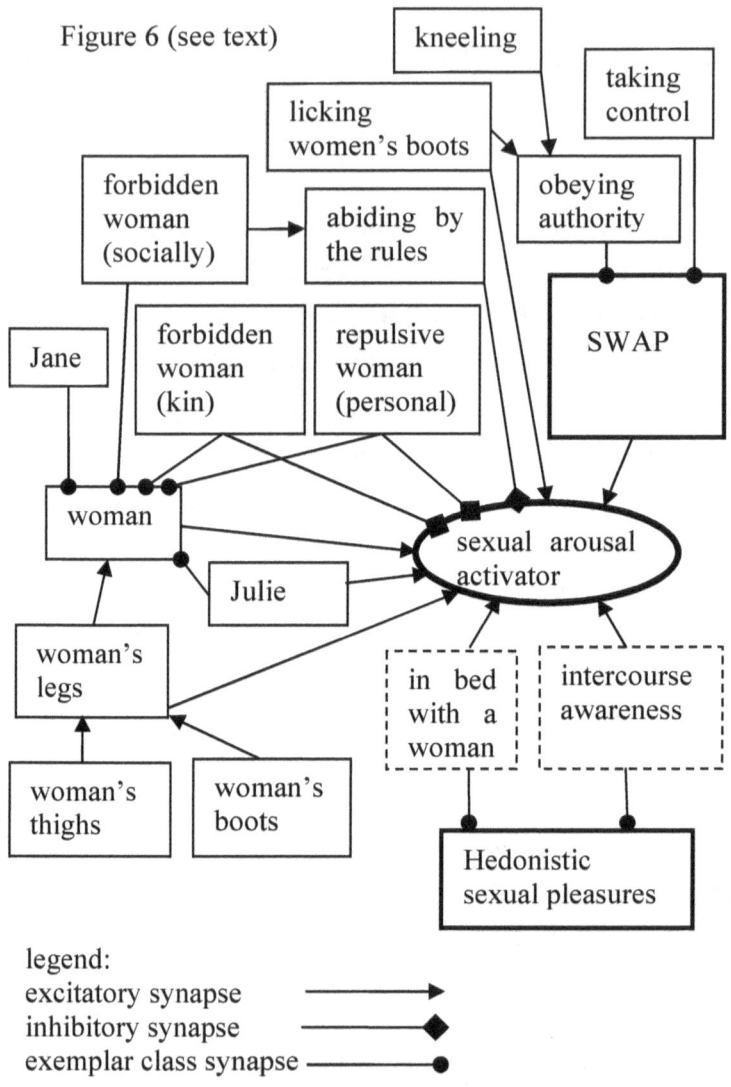

legend:
excitatory synapse
inhibitory synapse
exemplar class synapse

The heavy bordered frames denote innate nodes. They include the node that activates the physical mechanisms of arousal (the ellipse in the figure), and the two nodes (SWAP and Hedonistic sexual pleasures) that have guided the acquisition of Ramon's arousal cues. All the other nodes represent acquired concepts. The 'feeling protected while vulnerable' (SWAP) node has been active before and after puberty, and contributed to the acquisition of all the excitatory concepts, except those denoted by a dotted frame.

The concepts of the dotted frames were acquired during actual sexual experiences after puberty, and their formation was guided by the node 'Hedonistic sexual pleasure'. A major group of the latter is labeled as 'intercourse awareness'. During sexual intercourse, erotic physical centers, such as those in the genitalia, are stimulated and trigger the sensation of sexual pleasure. In addition to that, the participants are aware of a wide range of associated visual, tactile and auditory stimuli. Some of those, such as typical patterns of moaning and groaning, pelvic movements, and many others, become arousal cues by learning.

The inhibitory concepts could be learned without the guidance of the SWAP node. For example, learning incest avoidance may be through an innate learning activator that guides the formation of the inhibitory synapse from the node representing any person with whom the child grows.

The weights of the excitatory synapses are such that support from other firing nodes would be needed to activate the postsynaptic node. For example, the sight of a woman's boot will create sexual arousal only if it is on a woman's leg in the context of a sexual situation. That information would be provided by other firing nodes (not shown in the figure).

An exemplar-class synapse is a strong excitatory synapse and its relayed signal will activate by itself the postsynaptic node. For example, whenever 'Julie' is present, the node 'woman' is activated.

Some nodes, such as 'Julie', have direct connection to the sexual arousal node, in addition to their connections with other nodes. These are 'short-cut' connections. They evolve between nodes that fire together many times. The node 'licking woman's boot', which now has a direct connection to the sexual arousal node, was initially associated with the 'obeying authority' node. That connection may still exist, and the sensation SWAP would be triggered together with the sexual arousal.

All those nodes and the connections between them were shaped through standard neural learning mechanisms such as conditioning and generalization. The same mechanisms form all the other information structures in the brain.

Claire

Are Julie's information structures similar to Ramon's?

Morie

They are very similar. The only difference is in the actual data that they store. Major parts of Julie's information structure were formed through the guidance of the node 'SWAP'. Her experiences taught her that one way of achieving that feeling would be by taking control of intimidating situations. Ramon's actions stroke that cord in her information structure, and she responded by arousal.

Claire

Does Julie get aroused only by domineering men?

Morie

Not necessarily. When she grew up, she acquired cues for feeling SWAP. Those cues depended on her experiences. Now, having sex as an adult, being controlled by her partner may arouse her, as long as it triggers in her the feeling of SWAP. She may be aroused also by intermittent feeling of controlling her partner and being controlled by him. For example, when they roll in bed, the body language conveyed by who is on top sends corresponding messages of who is in control. However, both being on top or on bottom may trigger the feeling of SWAP. In other situations, she may feel SWAP by sheer domineering.

Bob

I find it interesting that in all our discussions, the entire arousal process was mental. You did not mention stimulating erotic points in the body. What about physical, as opposed to mental sexual arousal?

Morie

Physical and mental arousals go hand in hand. To really enjoy sex, the participants have to know the buttons that trigger both physical and mental arousal in themselves and in their partners. Sometimes, it may not be simple to uncover and to accept the mental buttons of yourself and of your partner, but it is worth the effort.

SIMILARITIES IN INFORMATION STRUCTURES OF SEXUAL BEHAVIORS

Bob

We said that Ramon is aroused by licking women's boots. Are there any similarities between this fetish and oral sex?

Morie

There are many similarities. Ramon uses social tokens of subordination to express his non-belligerence. The message is conveyed to Julie, who understands these universal symbols and is aroused by them. Similarly, social tokens are also used in oral sex to convey messages between the partners. Unlike the boot in Ramon's case, the human genitalia are ambivalent tokens, and they may be used to convey a variety of messages.

The penis is used as a symbol of a whole spectrum of attributes such as masculinity, strength, the most precious part of the man's body and the essence of being a man. It also symbolizes the mean, uncompassionate, brutish and degrading provinces of the male's soul. Similar connotations are associated also with female genitalia. All those connotations are widely known to all adults in any society, and any connotation may be chosen to convey a message during oral sex. For example, in fellatio, the implied message sent by the woman may be summarized by a statement like:

(a) I am holding your most precious part, you are under my control and I can make you feel whatever I want; I'll finish you.

(b) I am handling your most private part, I appreciate your trust, and I'll do my best to make you feel good.

(c) You are strong and masculine, and I admire it.

(d) I do what you want. With my pure lips, I satisfy your dirty desires. You are everything, I am nothing.

(e) I make you think that you are something special. What a fool you are! And many more.

The common to all such messages is their emotional meaning about feeling SWAP. If both partners are aroused by the ongoing physical actions, they give it one of those interpretations, or something similar. Each of them may assign a different interpretation to the same physical activity. They are aroused because their perception triggers the central emotion of 'SWAP'.

In this sense, there are similarities between Ramon's fetish and oral sex. However, there is one main difference. Unlike human genitalia, a boot does not have erotic centers. The only arousal that the boot creates is mental, whereas oral sex blends physical and mental ingredients.

Claire

Does feeling SWAP play any role in pure kissing and in "vanilla sex" i.e. no oral sex, no sex games, no fetishes; just plain "missionary" intercourse?

Morie

Anticipation and uncertainty are two of the emotions that novices feel as they embark on their first sexual interactions – be it a kiss or an intercourse. A mental target of the participants, who are already in a sexual state-of-mind, is to feel SWAP.

After a number of gratifying experiences, a novice may acquire new cues and develop arousal activity-plans that do not depend on feeling SWAP. Those cues are extracted from the details of the experienced sexual activity, due to their association with the sensation of pleasure that was created.

For example, before his first kiss, the boy is aware that he is crossing into the domain of the girl. He may be unsure about her reaction. He would like her to accept him, and it would be embarrassing if she does not. He needs to feel SWAP. They kiss and hug, relishing the new experience. Eventually, the stress component of his feelings subsides. His brain records bits and pieces of stimuli that he experienced during their activities, such as close-ups of her facial features, the warmth of her lips and the softness of her skin. Later on, those sensations, or even the thoughts about them, would arouse him without invoking the feeling of SWAP. Direct connections have been formed between the nodes of the new cues and the center of sexual arousal.

In most intercourse scenarios, including 'vanilla sex', the mutual positions of the partners send implicit emotional messages. The participants weave them into their SWAP activity-plans, at least at the onset of a sexual relationship. With time, other cues are acquired, which may bypass the SWAP nodes, and create pure hedonistic arousal activity-plans.

Bob

How similar is Ramon's information structure (Figure 6) to information structures in the brains of homosexuals?

Morie

Apparently, they are very similar. The same general processes guide the formation of those information structures. The only differences would be in the recorded experiences. If you replace the word woman with the word man in Figure 6, you will get an information structure of a gay man. If Ramon were a lesbian, her information structure would be the same as that of "Ramon the straight".

REVERSIBLE AND IRREVERSIBLE BEHAVIORS

Claire

The formation of the sexual arousal information structure takes place over many years, and it may never stop. How flexible is that structure? For example, can an existing information structure of a straight man be changed so that he becomes gay, or vice versa?

THE ROLE OF EXCITATORY AND INHIBITORY SYNAPSES

Morie

Information can be added to a structure so that existing behaviors change or stay the same. There are behaviors that can be modified and behaviors that are for good. It all depends. In general, a behavior is modified by modifying connections between nodes.

Whether a given behavior can or cannot be modified depends on the nature of the synapses that encode that behavior. It might be impossible to modify a behavior that is governed by strong, fixed, inhibitory synapses. For example, if Ramon's information structure contains an inhibitory, strong, fixed tie from the node 'man' to the node 'sexual arousal activator', Ramon will always be prevented from being sexually aroused by other men. It will be impossible to bypass that inhibition.

Sol

Will a similar strong, permanent, excitatory synapse guarantee that a certain behavior will always happen? For example, a strong tie from 'woman' to 'sexual arousal activator' will result in Ramon being aroused only by women.

Morie

There is a major difference between the two cases, and it has to do with a fundamental learning mechanism in neural networks that controls modification of synaptic weights. One fundamental mechanism can modify a synaptic weight only if both the pre-synaptic and the postsynaptic nodes have fired at the same time. We have seen this mechanism in action in Pavlov's conditioning. It guided the synaptic modification between the conditioned stimulus (the bell's sound) and the unconditioned response (salivation). The changes occurred because both nodes fired together (Figure 3). Now, imagine that there happens to be a fixed, strong, inhibitory connection between the node of the bell's sound and the salivation node. Whenever a bell sounds, the salivation node would be inhibited from firing. The fundamental learning mechanism will be unable to reverse this situation, because the muted salivation node cannot signal to the bell's node to increase the synaptic weight between them. Therefore, the dog could never salivate when a bell rings. This behavior would be irreversible.

Similarly, if in the brain of a gay man there is a strong, fixed, inhibitory synapse between the node 'woman' and the sexual arousal node, it would be impossible to replace that synapse by an excitatory one. The arousal node will never fire together with 'woman', and thus it will not signal to 'woman' to establish an excitatory synapse between the two of them.

On the other hand, in the brain of a gay man, a strong, fixed, excitatory synapse from the node 'man' to the sexual arousal node, will not prevent the establishing of an excitatory synapse from 'woman' to the sexual arousal node.

PLASTICITY

Morie

In general, if a certain behavior pattern is encoded by plastic ties, that behavior could be modified by exposing the individual to counter-experiences. For example, in Pavlovian conditioning, the learned response to the bell's sound could be reversed by exposing the dog to rings with no food. This is called extinction. Similarly, a man who is attracted to a certain woman and later finds her repulsive would be able to detach from her.

Human learning of crucial life behaviors is often subjected to fixation by imprinting (like that of goslings after hatching). It is a fact that certain human behaviors can be learned only within well-defined time windows. After that time window has elapsed, those behaviors cannot be changed. It is possible that the neurons that activate those behaviors lose their plasticity, and all their synapses become fixated. If such neurons participate in sexual attraction and arousal, then certain aspects of those behaviors could not be modified.

For example, in Ramon's case, if the node 'sexual arousal activator' loses its plasticity when only the information of Figure 6 is recorded, new nodes will not be able to develop direct connections to 'sexual arousal activator'. Ramon will never be aroused by forbidden women (incest and repulsive), because this information is imprinted and cannot be bypassed. On the other hand, if social rules change, say by changing the minimal legal age of engaging in sex, Ramon will be able to adjust. This information is not represented by the non-elastic synapses on 'sexual arousal activator'. In Figure 6, that information is encoded by synapses of the node 'abiding by the rules'.

If in addition to the node 'sexual arousal activator' other nodes lose their plasticity, other behaviors become fixated. For example, if in Figure 6 the node 'SWAP' loses its plasticity, no new emotional stimulation patterns of this kind could be learned. If the only exemplar of that class is 'being subordinate', then Ramon will be unable to acquire cues that arouse him as 'being dominant'. However, that will not affect the learning ability guided by 'sexual pleasure'. New cues, such as sexual positions, that are generated in the course of intercourse could be acquired.

Also, the information structure can always be expanded when non fundamental entries are involved. For example, new features of 'Julie' may be learned and become new arousal cues.

It is not known yet which neurons encode which particular aspects of sexual behavior, let alone if and when they become fixated. It is also not known to what extent the level of fixation is universal, or whether it varies from one individual to another. Circumstantial evidence suggests that fundamental features of sexual attraction and arousal become fixated, in many cases before puberty. In numerous cases, that fixation seems to be so strong that it could not be modified or bypassed by the willpower of the individual or by social dictates that are backed up by severe repercussions.

According to the model, all those aspects of human behavior could be explained by general fundamental biological processes, which are not limited to sexual behavior. Some fundamental processes result in fixated behaviors, while others result in modifiable behaviors. Only when the specifics of the involved processes are uncovered, the question of the constancy of sexual orientation could be answered with scientific accuracy and specificity.

FEELINGS AND ACTIONS

Sol

So far, we have discussed what information triggers the feelings of sexual arousal and attraction. How do these feelings turn into actions?

Morie

Those feelings, like any other thoughts or feelings, turn into actions through thinking processes. We will talk about thinking processes in more detail later, but for now, it might be interesting to observe only the first steps of the realization of those feelings.

In order to procreate, a man and a woman have to make contact with each other. To accomplish it, the sexual attraction and arousal nodes trigger in the brain a request to get closer to the potential mate. That request would be thought over and evaluated by the brain, and if an approved plan is found, it will be carried out. After they get close to each other, an instinctive activity-plan to explore the partner would be triggered in their brains.

The plan to explore objects is one of the earliest plans of the brain. In a self-rewarding instinct, newborns bring objects to their mouths. Once the object gets to the mouth, the baby is transformed to a state of blissful existence, while the brain is processing the inflow of sensory information. Later in life, this instinct is replaced by a need to touch. Both instincts are in full operation at the first stages of sexual activity: the participants feel an urge to touch each other. It is interesting that now, in the adult, the otherwise abandoned urge to touch and explore with the mouth is as strong as the urge to touch with the hands. Maybe, the neural circuitry that controlled the oral urge in the baby has not been discarded after all. It is being used in the adult to produce the oral sexual urge. The resulting infantile sensation of fulfillment has matured in the adult into sexual oral gratification.

SUMMARY

Claire

We have seen examples that illustrate how the learning of sexual arousal and attraction cues could be explained by a few biologically feasible principles and mechanisms. Those are the core principles, which guide the development of all sexual behavior patterns. The wide variety of sexual behavior pattern that we see is the result of the combinations of those mechanisms with individual experiences and genetic predispositions.

In order to be able to learn the sexual cues, the system needs innate detectors that can point to them. Those detectors identify events from which the cues have to be extracted. The detectors could be divided into two main groups: sensual detectors and emotional detectors.

Sensual stimuli trigger sensual detectors that guide the system in the acquisition of cues that were associated with those stimuli. Those detectors include erotic sensors, found mostly in the genitalia, which generate the sensation of sexual pleasure. Learning through these sensors is most common after the onset of puberty.

In addition, detectors that identify the sex of the partner are important in establishing the target of sexual attraction. In many species, this detection is done by pheromones, which are genetically designed to function in both sexes. It is not yet known what takes the place of pheromones in humans. It is suggested that the pitch of human voice may serve that purpose. However, this has not been scientifically proven or refuted. When used in our discussions, it should be treated as a generic term for a stimulus that differentiates its source by its sex, and that triggers different responses, depending on the sex of the receiver. These detectors could be active consciously or unconsciously throughout childhood.

Emotional detectors have the same role as sensual detectors, but they are triggered by emotions. It is suggested here that the combination of two universal innate emotions, the feelings of safety and vulnerability, is the trigger for learning sexual cues. There is ample circumstantial evidence of the participation of those emotions in establishing those behaviors. The role of other emotions cannot be ruled out. Emotion detectors are active before and after the onset of puberty.

All learning is realized by representing concepts by nodes and encoding relationships between concepts by synaptic weights. Information is represented as interconnected class-exemplars and item-parts hierarchies. The processes of recruiting nodes for establishing new concepts and modifying synaptic weights for establishing their relationships follow localized rules that depend on the firing state of the involved nodes. Processes, such as classical conditioning (Pavlovian learning) and generalization, which are common throughout the entire neural system, take part also in the acquisition of sexual cues and the construction of the information structures in which they operate.

An important aspect of learning sexual cues is imprinting. Many learning processes can occur only within preset time windows. Information learned within such time windows cannot be erased afterwards. The implication for sexual cues is that some of them become fixated in the system, and are difficult or practically impossible to modify.

CHAPTER THREE

HIGHER MENTAL FUNCTIONS

REFLEXES AND THINKING

Claire

I got the impression from the relationship between Roman's sexual behavior and his information structure, as described in Figure 6, that his behavior consists of many reflexive-type activities. A stimulus activates a node, which starts a chain of activation of other nodes, the last of which is arousal. Does thinking contribute to Roman's sexual behavior?

Morie

Thinking is a wide-ranging brain activity that consists of many sub-processes. By analyzing thinking, it is possible to identify the basic mechanisms and the sub-processes that contribute to what we experience as thinking. The details of the underlying biological mechanisms, though, are for the most part unknown. Therefore, we will treat the brain as a thinking apparatus that consists of input channels, to which the givens are attached, and output channels, at which the solutions could be found. The information structures of the brain are accessible to its thinking apparatus. In thinking, basic mechanisms modify existing information structures, and guide the firing of nodes. We will describe how some advanced thinking processes are built from a few basic thinking mechanisms. Let us start by comparing reflexive and thinking processes.

INFORMATION RETRIEVAL

A main sub-process of thinking is information retrieval. In thinking, active information entities trigger other information entities, thus retrieve them. This happens also in reflexes. However, there are two main differences between reflexes and thinking. The first difference is in the retrieval mechanisms; those of thinking are more elaborate. The second difference is in the nature of the retrieved information. Reflexes retrieve old information, while thinking may generate new information from old one. Here are some examples.

Paul usually recognizes his friends immediately as he sees them. The sensory input of a friend stimulates in Paul's brain the node that represents the friend. This is, basically, a reflexive process, much like blinking your eyes when something moves suddenly in front of you. The sensory input of the moving object triggers the nodes that activate the muscles of the eyelids.

When Paul attended his thirty-year class reunion, a beaming, middle aged, zaftig woman approached him from across the room. "Look, look, look, she said. Isn't that Paul?" Paul looked at her and felt uneasy. Who is that woman? After a long pause, it clicked: she is the vivacious, skinny cheerleader from back then. "Mary! You have not changed a bit!" he exclaimed. The combination of cues that Paul's brain was getting could not at first trigger the node representing the cheerleader. He had to focus his mind on specific features such as her smile, her tiny nose and her voice, while ignoring other signals, such as her waistline. Only then, the node representing her fired in his brain.

MODULATION IN DATA RETRIEVAL

The mechanisms of sifting through data, ignoring some parts of it and enhancing others are typical to thinking processes. They distinguish thinking from reflexive processes.

Modulation could be the biological mechanism underlying ignoring some parts of the data and enhancing others. In modulation, a node whose role is to manipulate data affects the flow of signals between two other nodes. Signals from the modulating node weaken or strengthen signals that are exchanged between the two modulated nodes that represent the data. One plausible way is that the modulating node temporarily strengthens or weakens the synaptic weights between the involved data nodes.

The modulating node in Paul's brain strengthened signals to nodes that represent Mary's smile, and weakened signals to nodes that represent her size. As a result, the node that represents Mary the cheerleader was activated.

COMBINATION OF BASIC RETRIEVALS

Another difference between reflexes and thinking is in the complexity of their retrieval instructions. Retrieval instructions in thinking can be much more complicated than in reflexes. For example, consider the question: "Fred's father is the brother of Mike's mother. What is the relationship between Fred and Mike?" It will not be too difficult to retrieve the answer that Fred and Mike are cousins. In this case, 'cousins', which is a node in the memory, was retrieved by a non-trivial retrieval request.

RECRUITMENT OF NEW NODES

A third difference between reflexive and thinking processes is that the retrieval process in thinking may add new nodes to the existing information structures.

For example, in the late sixteenth century, William Bourne, an English innkeeper, asked himself the question: "how to make a boat that can go under the water and then float again?" In his brain, this question was expressed as a complex retrieval request. The goal of the retrieval request was a node that satisfies the given cues. However, such a node did not exist in his brain. Bourne knew about boats and about Archimedes principle, which deals with the buoyancy force. The un-answered retrieval request turned in his mind into a creative thinking goal. He thought about it and was able to come up with a plan for a submarine, which did not exist before that in his brain or elsewhere. The plan itself was expressed in his brain as connections between new nodes and existing information structures. (Incidentally, Bourne's submarine was never built. The first submarine that sank a warship was the CSS H. L. Hunley in 1864).

The process of creative thinking is very common. Quite often, the brain invokes it automatically when an information retrieval request cannot be met in a satisfactory way.

Creative thinking is responsible, for better and for worse, for all the innovations, big and small, in the arts and sciences and in all other walks of life that affect humankind. Creative thinking is also indispensable in solving innumerable, daily, personal, mundane problems.

Claire

If thinking processes are so different from reflexive processes, how are they carried out by the brain?

RETRIEVAL REQUESTS

BASIC RETRIEVAL REQUESTS

Morie

Retrieval of information is a central part of thinking processes. Although we do not know their biological details, we may glean some knowledge of the basic retrieval processes by analyzing the gross features of thinking.

Since information is organized in the brain as associative items-parts and classes-exemplars hierarchies, basic retrieval requests have to accommodate that organization. Therefore, four basic retrieval requests that the brain uses should be:

(1) Retrieve an item of a given part.
(2) Retrieve a part of a given item.
(3) Retrieve a class of a given exemplar.
(4) Retrieve an exemplar of a given class.

Examples: When we feel hungry, we think about food. Our first thought is what to eat. That translates into a basic retrieval request to retrieve from our memory an exemplar of the class 'food'. That is denoted as: retrieve [an exemplar of 'food']. (Square brackets will indicate in the following a basic retrieval request.) The outcome, depending on the individual, could be 'pizza', 'salad', 'bread'. etc.

The question "what is a pummelo?" is a retrieval request: retrieve [the class of 'pummelo']. The outcome depends on the individual. The retrieval mechanism of those who do not have the information in their system would come up with 'nothing found', prompting the answer "I don't know". The retrieval mechanism in those who have the information would prompt the answer "a citrus".

APPRAISAL OF RETRIEVED INFORMATION

As the last examples illustrate, a retrieval request may result in one retrieved entity, in a number of retrieved entities, or in no retrieved entity at all. The system must have means of appraising the nature of the outcome. An appraisal unit would notify the rest of the thinking apparatus whether one, none, or many entities were retrieved, so that the next appropriate step of the thinking process could begin.

LOGICAL COMBINATION OF RETRIEVAL REQUESTS

By using the logical operators AND, OR, NOT, and delimiters, basic retrieval requests may be combined into compound requests.

Claire

What are delimiters?

Morie

In our case, delimiters are markers that mark the beginning or the end of a part of a compound request. In the following, we will denote those delimiters by parentheses.

Bob

Why do we need delimiters in retrieval requests?

Sol

Let me give you an example. At lunchtime, one of your office mates goes out to bring lunch for every one. You tell him that you would like to have "a burger and onion-rings or French-fries". He will probably get you what you wanted. If you give a robot the same instruction, you may end up with a bag of French-fries.

Bob

Why?

Sol

In your request, you did not spell out the delimiters. Your friend understood that you wanted a burger AND (onion-rings OR French-fries)}, meaning (a burger AND onion-rings), OR (a burger AND French-fries). The robot, depending on how it was programmed, interpreted your request as (a burger AND onion-rings) OR (French-fries). He fulfilled your request by bringing French-fries.

NESTED RETRIEVAL REQUESTS

Morie

The four basic retrieval requests retrieve a node based on its relationship with another given node. This is the simplest case. The brain can handle also nested retrieval requests. In a nested request, a second retrieval request replaces the given of the basic request. For example, the basic request: retrieve [an exemplar of a 'given item'] could be expanded to: retrieve [an exemplar of [an item of a 'given part']]. The second request is "nested" in the first one. The second request may be a basic request, or it may nest in it other requests.

For example, the request "name an animal that has wings" is translated into: retrieve [an exemplar of [an item of 'wings']]. The nested request [an item of 'wings'] would retrieve 'bird', because 'bird' is an item whose part is 'wings'. After this step, the original retrieval becomes: retrieve [an exemplar of 'bird'], which may result in any or many bird names.

EXAMPLE

Becky

Could you please give a non-trivial example of information retrieval, and show how it could be accomplished when information is represented as a network of nodes?

Morie

OK. Let us see an example that illustrates the role of modulation in information retrieval. Consider asking Beth to name an actor who played in an ancient movie. The relevant parts of her information structure are shown in Figure 7.

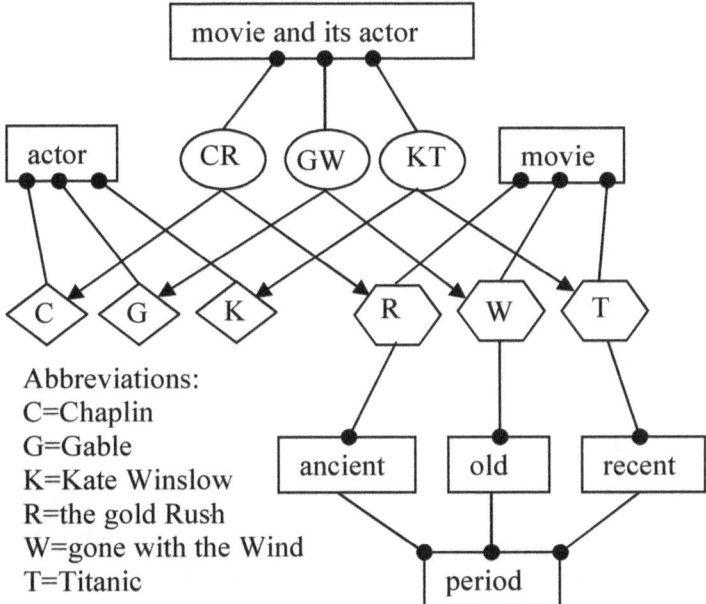

Figure 7 An example of an information structure. Circle head arrow indicates a connection from an exemplar to a class. Pointed arrow indicates a connection from an item to a part.

The figure shows how information about three movies and their actors is stored in her memory. Hexagons indicate movies, diamonds indicate actors, and circles indicate combinations of a movie and its actor. Rectangles indicate classes. The first movie is The Gold Rush (R), with Charlie Chaplin (C). The second movie is Gone with the Wind (W), with Clark Gable (G). The third movie is Titanic (T) with Kate Winslet (K). A circled-head arrow indicates a connection between an exemplar and a class e.g. between Chaplin (C) and the class 'actor'. A pointed arrow indicates a connection from an item to its parts e.g. the item CR ('Chaplin in The Gold Rush') has the parts 'Chaplin' and 'The Gold Rush'.

In general terms, Beth's assignment is to retrieve an information entity based on its specific relations with other given information entities.

(When expressed formally by retrieval requests, the assignment could be stated as: retrieve [*an exemplar of* 'actor'] AND ([*a part of* [*an exemplar of* 'movie and its actor']] AND ([*an item of* [*an exemplar of* movie] AND [*an exemplar of* ancient])). As you can see, this compound retrieval request has nested requests, logical operators and delimiters.)

Figures 8A-D are four snapshots of the retrieval process. Highlighted nodes indicate firing nodes. The figures show how activation propagates in the network. The retrieval starts by activating the nodes that represent the givens (Figure 8A). The thinking apparatus then modulates synapses from firing nodes until downstream nodes with the required association to the firing nodes are activated. Activation propagates in the network until the target node fires (Figure 8D).

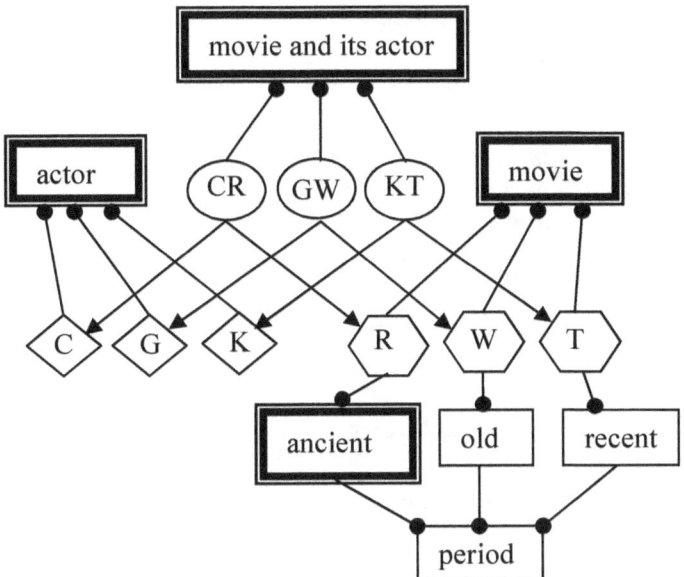

Figure 8A The nodes that represent the givens of the assignment are activated. The synaptic weights of their outputs are modulated to zero, so no other node is active.

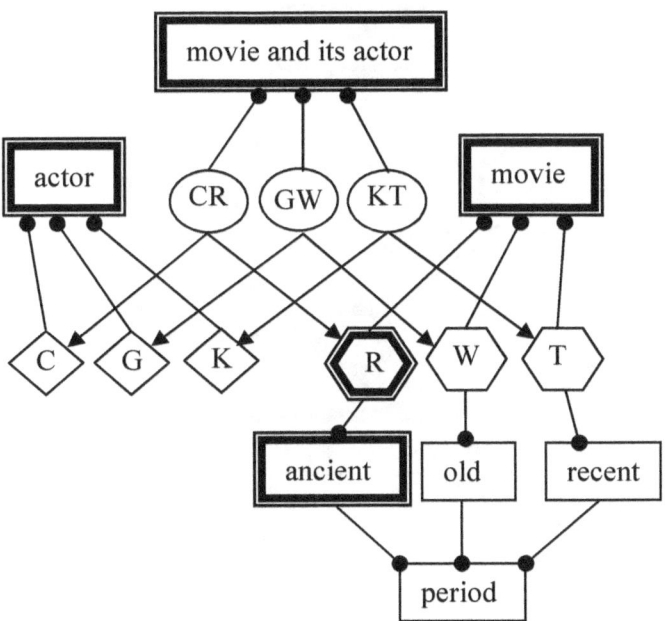

Figure 8B Modulation starts. The synaptic weights of the givens are modulated gradually, until the node "R" (the gold rush) fires due to signals from 'ancient' and 'movie'.

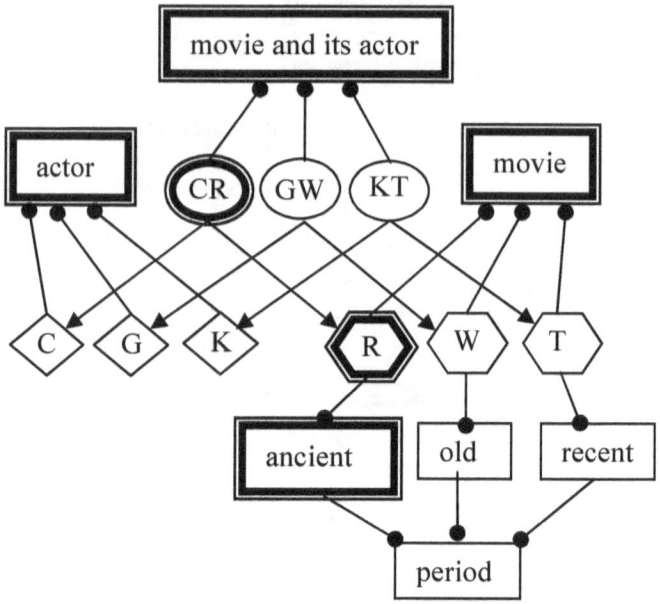

Figure 8C The synapses of the givens and of "R" are modulated from zero up, until a new node, "CR" (Chaplin in the Gold Rush) fires.

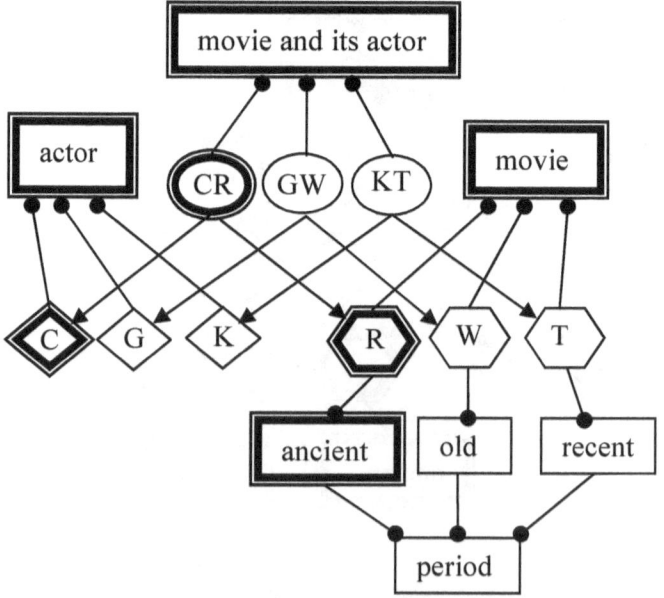

Figure 8D The synapses of the givens, "R", and "CR" are modulated from zero up, until a new node, "C" (Chaplin), fires. The retrieval request is completely satisfied.

The thinking apparatus oversees the retrieval process. It modulates the weights of the output synapses of firing nodes. It first modulates the weight to zero, and then increases the weight until a new downstream node fires. It then appraises the outcome. If a satisfactory node has not been found yet, the search continues, and the output synapses of the newly firing nodes are modulated until their downstream nodes fire, and so on. When a satisfactory solution is retrieved, the search ends.

Sol

Do all the thinking retrievals in the brain use the same retrieval strategy?

Morie

No. Different situations could arise in the process, and the thinking apparatus should have mechanisms to handle them. For example, if a retrieval search seems to lead to nowhere, or if it is stuck before reaching a solution, or the retrieved nodes satisfy only parts of the request or contradicts parts of it, then the thinking apparatus should be able to take correction measures. Correction measures include starting another search without some of the nodes that might have caused the problem, or inhibiting some of the retrieved nodes if too many of them are active, and so on. Figure 8 just illustrates how a compound retrieval request could be resolved into basic retrieval requests, joined by logic operators.

Expressing a compound request as a combination of basic requests makes it possible to retrieve information from classes that are not represented by a node. The class 'actor who played in an ancient movie' does not exist in Beth's memory, yet, an exemplar of it could be retrieved.

Micropsychology

Claire

In this example, Beth's memory recorded only one actor of The Gold Rush. What would be the outcome of the thinking retrieval if two actors had been recorded in her memory?

Morie

It would depend on the details of the process, and the existing synaptic weights. Both actors together might have been retrieved first. Then the system could scan one actor after the other, by modulating down the synaptic weights that lead to their nodes.

INFORMATION AND ACTIVITY-PLANS

Sol

The examples that we discussed so far were about thinking tasks that involve only information retrieval. How stored information is retrieved and used in thinking tasks that guide and control actions?

Morie

Sensory concepts, internal feelings, and physical activities are all represented in the brain by nodes. In order to participate in a physical or mental activity, the involved nodes have to fire. Similar to an emotion, which is felt when its node fires, or to a sensory concept, which is sensed when its node fires, an activity is executed when its node fires.

However, it seems that many concepts have several representations in the brain. A sensory concept is represented by the sensory node that detects it, but it is also represented by the node that represents the word that describes it, and probably by other nodes that enable us to think about it in a non-verbal way.

Similarly, physical activity is represented by a node that actually activates the muscle unit, and by "deeper" nodes, that are used by the brain for planning that activity. Before an activity is carried out, its deeper representation is evaluated. If it is deemed beneficial, it is cleared and it is executed by the "superficial" nodes that activate the motor units of the body.

Often, activities are executed according to retrieved activity-plans, which are memory records that were compiled based on previous experiences.

Becky

Can we see an example of an activity-plan in action?

Morie

Consider a common activity: walking to the corner of the street, watching the traffic lights, and crossing the street. There should be an activity-plan in our memory for such a routine. One possible activity-plan is described in Figure 9.

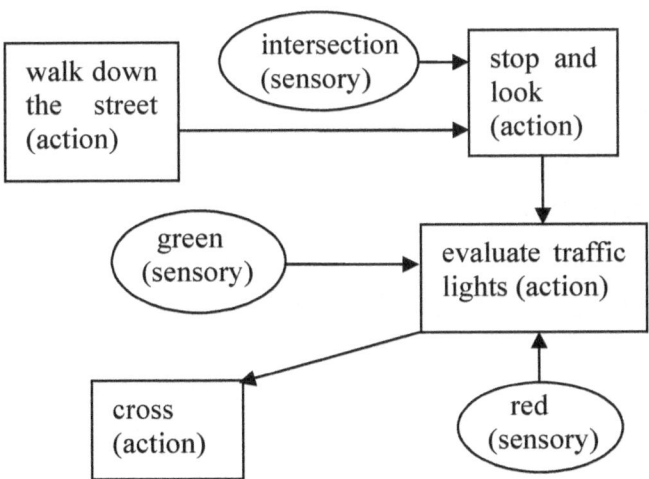

Figure 9 An example of an activity-plan

Nodes that control activities are denoted by rectangles, and nodes that receive and process external stimuli are denoted by circles. All those nodes are class nodes. The sensory nodes are activated by any exemplar of their class. For example, the node 'intersection' is activated when we receive stimuli from an intersection. Nodes that represent actions activate exemplars of their class, according to the actual situation. For example, the node 'walking down the street' activates all the needed walking routines.

Sol

Why are the nodes of the activity-plan class nodes and not item nodes?

Morie

That makes the activity-plan general. Had it been based on item nodes, it could be used only to repeat exact past events, for example, walking only in one specific street.

Figure 9 describes the sequence of activated nodes, starting as we walk down the street. When we come to an intersection, the stimuli that we receive combine with those of our walking. That activates the node that makes us stop and look around. When traffic lights are identified, they become the focus of our attention. The instruction to cross the street is activated when the light is green.

An activity-plan performs two kinds of information processes: classification and retrieval. The incoming sensory information is classified by the detectors of the activity-plan. For example, traffic lights are identified as such and are represented by the appropriate firing node, e.g. a node for a red traffic light. The node that represents the specific traffic light becomes an exemplar of the class 'red traffic light', which is part of the activity-plan. The responses are derived by information retrieval requests that merge classified sensory information with information about actions. The basic retrieval instructions are the same as in other information retrievals e.g. retrieve an exemplar of a given class (walk down the specific street), retrieve an item ('stop and look') whose parts ('walking in a street' and 'reaching an intersection') are given, etc.

MENTAL SIMULATIONS

Morie

An important sub-process of thinking is mental simulation. In mental simulation, nodes that represent actual events are activated by brain's internal mechanisms, without the events really occurring. The activated nodes may represent parts of actual events that occurred before, or combinations of events that have never occurred together. For example, we may think about whom we met today at work, or about our meeting next week with a new boss. Usually, such thoughts do not activate complete and accurate picture representations of the events. Rather, they activate nodes that represent the general situation, without all the details. Detectors detect the features of the activated events and relay that information to other parts of the brain that respond to them. The responses may be emotional, actual physical activities, and incorporation in the information structures of the brain. Thus, the simulated event and all its associated emotions and perceptions are experienced by the brain. Mental simulation is common in creative thinking. It enables the brain to retrieve information from imaginary events, and create new outcomes based on combinations of old experiences.

VERBAL THINKING

Claire

Can we look into an example of the use of mental simulation in creative thinking?

Morie

Imagine Shakespeare working on his celebrated opening statement of Hamlet's soliloquy. We don't know what went on in the Bard's mind, but it is safe to assume that he was using words and language elements that were recorded in the Wernicke's and Broca's areas of his brain, and which had connections to other brain centers. He retrieved words and phrases, analyzed them, and weaved them into the plot.

Shakespeare apparently wanted the opening statement to allude to the passage that would follow. The passage itself lists various sufferings that are common in life: "slings and arrows of outrageous fortune, a sea of trouble, heart-ache, the thousand natural shocks that flesh is heir to, the whips and scorn of time, the oppressor's wrong, the proud man's contumely, the pangs of despised love", and more. On the other hand, the afterlife is an unknown country from which "no traveler returns", and it may have sufferings of its own. Hamlet considers these information items as he weighs the advantages and disadvantages of committing suicide. The retrieval request in Shakespeare's mind could be formulated as: find class nodes that expresses Hamlet's dilemma. All those sufferings are exemplars of one class node. The exemplars of a second class node convey the doubts about suicide. He might have come up with a phrase like "life is full of suffering, is suicide a solution to all life's problems?" Let's call this phrase 'version one', for brevity.

Shakespeare wants to gauge the effects of this opening statement on an imaginary audience. He enacts in his mind a mental simulation in which he is in the audience, and 'version one' followed by the rest of the passage is played to him. He feels the emotions that the passage stirs in his mind. The 'boring' detector fires. Shakespeare feels bored. The opening statement divulges too much. The passage repeats the idea that was already fully expressed by 'version one'.

Shakespeare changes 'version one' to 'version two': "to be alive or not to be alive: which is better?" He enacts a second mental simulation. This time it does not trip the 'boring' detector, but another one fires. A detector that Shakespeare adopted recently – "brevity is the soul of wit". 'Version two' is too wordy. He tries in his mind some trimmed versions of it and he comes up with "to be or not to be: that is the question". Halleluiah! And the rest is history.

Claire

The scenario that you laid-out describes creative thinking as a completely rational process. Shakespeare knows exactly what to look for, and then he finds it. Isn't there a room in creative thinking for some gut-feelings and intuition?

Morie

Of course there is. But even intuition and gut-feeling decisions are processes in which information is retrieved and manipulated by the brain. All information is encoded in the same way in the brain's information structures. Therefore, it must be retrieved and manipulated by the same basic mechanisms.

The main difference between intuition and rational thinking is in the awareness of the thinker to the thinking process. In the previous example, Shakespeare was aware of his entire thinking process, and could provide the rationale for modifying the opening statement: first, it took away the element of surprise, and then, it was too long. However, he would have reached the same conclusion if his entire thinking process were done unconsciously. If so, he might not be able to pinpoint why he did not like 'version one' and 'version two', and how he arrived at the final version. He would attribute the entire process to his gut-feelings or to creative intuition.

Sol

Are mental simulations applied only in high-level thinking processes?

Morie

Not at all. Any thinking process whose outcome includes new combinations of old experiences has most likely passed through mental simulation. Evaluating personal relationships, planning daily activities, cooking a new recipe, and many more are just few examples of thinking processes that include mental simulations.

NON-VERBAL THINKING

Sol

In a mental simulation, we think about various aspects of the simulated event. How does the brain do it?

Morie

In the musical "Fiedler on the Roof", Tevye the Milkman imagines that he becomes a Rothschild. One of the things that he would do would be to build a big house with three staircases; one going up, one coming down, and one leading nowhere, just for show. Consider a closer-to-home example. Terry Lott, a next-door neighbor, buys a lottery ticket and imagines that he hits the jackpot. Figure 10 illustrates the simulation process in his mind, and the involved information structures. The node that represents himself becomes an exemplar of the class 'rich', and activates it. The class node 'rich' activates its inherited nodes 'owns chauffeured limousine' and 'owns a yacht'. Unlike Shakespeare's information structure in the previous example, whose nodes represented words, Terry's 'yacht' and 'chauffeured limousine' represent snippets of sensory information. He visualizes a shiny limousine and a uniformed chauffeur opening the door for him. He can smell the salty foam of the waves, as the yacht slices through the blue sea, under sunny, cloudless skies. These activated sensory images are detected by Terry's detectors and are processes by his network. They trigger various typed of nodes. He feels emotions such as happiness and pride; he feels the

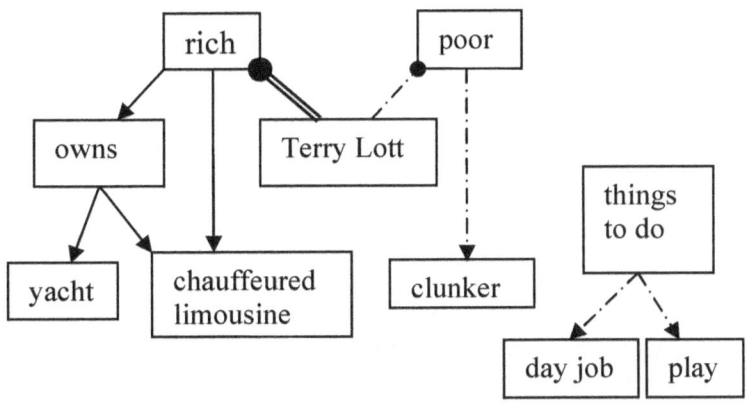

Legend: temporary connection that enacts the mental
simulation
Imagined information
Realistic information

Figure 10 Underlying information structures in a mental simulation.

pleasantness of breeze on his face in a balmy day; activity-plans are then triggered by those emotions and a blissful smile covers his face. Terry enjoys his fantasies.

Then, the node 'thing to do' fires. This node, which was activated by other brain centers, is not part of the ongoing simulation. Many independent processes can take place in the brain simultaneously. Terry has been awake and idling for a while. Nodes in his system monitor these aspects of his life and make him a responsible person. They activate nodes that remind him that he has a day job. Another node reminds him that his car, the old clunker, needs repairs. Various activity-plans, with conflicting goals, are active now in his mind. A control center in his brain notices that situation, and decides that Terry has been fantasizing enough, and it is time to get back to reality. The node 'rich' is disconnected from the node 'Terry Lott', and all his fantasies fade away.

Bob

I see the similarities between the mental simulation processes in those two examples. Do similar processes underlie fantasies during sexual interaction and masturbation?

Morie

Yes. The difference is in the contents of the information and where this information is stored in the brain, but the processes are the same.

Becky

If the processes are so similar and so natural, then why President Clinton's health secretary, Dr. Joycelyn Elders, had to resign her post after she suggested in a United Nations conference on AIDS that promoting masturbation might be an appropriate means to discourage young people from engaging in riskier forms of sexual activity?

Morie

Her idea triggered a political maelstrom. I cannot tell you what exactly went on in the politicians' minds. Cold statistics suggest that most of those who opposed her idea have enjoyed that same activity themselves, and apparently no one was hurt. However, they preferred to hold on to their traditional value systems.

DYNAMIC RETRIEVAL MODIFICATIONS

Morie

The brain solves elaborate retrieval requests by dividing them into smaller pieces, and processing piece by piece. It evaluates each partial outcome, and if it is not satisfactory, the brain may modify the original request, until a satisfactory outcome is obtained. The brain has circuitries that are responsible for carrying out such tasks. The details of those circuitries are not known, but we can get a glimpse on their dynamics by drawing parallels between them and how we retrieve information from external sources.

Consider the following example. Grandma wants to visit her grandchildren during Labor Day weekend. She decides to buy the tickets online. She enters the information by filling in the blanks, and the computer compiles them into a compound retrieval request, consisting of standard basic requests. Her request specifies the departure and arrival airports, the dates, and other required information. After a short time, a list of many flights, which satisfy her request, appears on the screen. The request activated many nodes that qualify as solutions. Grandma notices that the flights have different prices. That triggers in her mind the activity-plan that looks for the lowest price. As the least expensive flight is retrieved, the detector that indicates 'cannot afford it' fires. Her original search, which retrieved many possible outcomes (many firing nodes), was trimmed by an additional constraint, and now nothing has been retrieved (no node is firing). In response, Grandma now modifies some of the original basic requests. She has flexibility in the dates, so she tries other dates. The original dates, which are exemplars of the class 'date', are replaced by other dates, which are also exemplars of that class. The new retrieval produces a number of flights. The cheapest, a red eye with a change of planes, is affordable. Grandma simulates the situation in her mind. She sees herself sitting for hours in a waiting lounge, tired and eager to be with her grandchildren. The node that represents the emotion of feeling unhappy fires, and triggers the activity-plan of 'continue the search'. Nodes that represent other flights are triggered, and an affordable and convenient flight is finally found.

In this example, the brain modified the original request due to unsatisfactory outcomes. Replacing one exemplar by another exemplar of the same class, or one part by another part of the same item, creates a modified request that is similar to the original one. The brain uses this common strategy when a search does not produce a satisfactory outcome. The brain may also deviate from the original request by ignoring parts of it, or by accepting solutions that are not completely satisfactory.

PERCEPTION

DETECTORS AND ACTIVATORS

Morie

From an information-processing perspective, a node may be viewed as an element that has two parts: a detector and an activator. The detector is exposed to an information stream. It sits in waiting, and once a specific information pattern passes by, it detects it. It then triggers the activator, which sends signals to other nodes.

There are specialized nodes whose role is to learn new information, and to incorporate it in the brain's information structures. The activators of those nodes manipulate synaptic weights of other nodes in order to accomplish their task. Such nodes are responsible for assembling parts into items, assigning items as exemplars of classes, and executing various learning paradigms such as classical and operant conditioning. The manipulation of the synaptic weights, which is how the learning is realized, is carried out by basic learning mechanisms that are activated by the activators.

For example, when we taste for the first time a sweet fruit, say a date, visual detectors recruit an item node 'date' to represent the visual parts of a date. The taste detector 'sweet' senses the sweetness, and regulates the formation of synaptic weights between 'date' and the class node 'sweet'. This establishes the new node 'date' as a part of the brain's information structure. If that person craves sweets, the node that regulates classical conditioning would establish connections that will make him salivate the moment he sees a plump date.

The input from the world enters the brain as data streams from pixels. The brain creates our perception of the world by generating two types of information about those streams: what items are there, and what is the significance of those items to us. An item consists of a pattern of pixels that are treated by the brain as one entity.

If a certain pattern of pixels repeats itself often, it might be of greater significant to the individual. Therefore, one of the innate mechanisms of the brain identifies repeating patterns of active pixels, and groups them together into items. Itemizing serves two purposes; first, new items develop associations with other items. Second, a node that represents an item serves as a detector of that item. Whenever the same pattern occurs again in a new data stream, it is immediately recognized. It is then simpler for other detectors to further analyze the information stream. Extraction of prevailing patterns in the information stream is one way by which the brain develops detectors that depend on the experiences of the individual.

For example, a baby defines items in her surroundings based on their prevalence. Soon she is able to distinguish between familiar and unfamiliar items. She then can associate familiar items with other familiar items, and expand her associative information structure, based on her experiences.

In addition to the innate mechanisms that itemize data streams and develop detectors based on prevalence, there are innate detectors that are triggered by certain pre-set patterns in the data stream, such as the presence of sugar. Those detectors, too, have dual roles. First, they notify other parts of the brain about the presence of the detected pattern. Second, they learn from their experience, and add new cues to the repertoire of their original cues. They expand the repertoire of their triggering cues.

For example, in Pavlov's experiments, innate food detectors, which are triggered by certain chemical compounds found in food, notify the neurons that control salivation about the presence of food. In addition, they contribute to the expansion of the innate food detector. After the conditioning, the food detector would respond not only to the pre-set chemical compounds, but also to the ring of the bell. Such expansions are expressed as logical combinations of cues.

Cues that trigger the detector are joined by the logic operators AND, OR, and NOT, and form a compound cue. For example, it might be possible to train a dog to salivate to 'real food' OR 'bell ring' OR ('a flash of light' AND 'a hand clap') AND NOT to ('a bell ring' AND 'a flash of light'). Each pattern in such logical combinations may become a detector on its own. A bell ring may become a distinct detector, triggered by a bell ring, notifying other parts of the brain when such a pattern has been detected. It then could also expand its own cue basis with other cues that have been associated with that ring. For example, if the dog notices that the trainer raises the bell before it rings, raising the bell may become a cue to the ring itself.

The innate brain has a set of basic detectors. Those detectors are expanded by experiences, much like the dog's food-detector that was expanded to include the bell ring. Together, basic and expanded detectors form a network that determines how we perceive the world.

All the detectors that our brain uses could be traced, in principle, to the basic detectors of the brain. All the detectors in our brain are the outcome of the basic detectors and their interactions with inputs from the environment.

Sol

When we are born, our brain stores very little information, but it has innate detectors and activators that regulate the acquisition and organization of new information. Those detectors and the information structures that their activators organize determine our perception of the world. Could you list all the basic detectors that determine the concepts that our brain could ever acquire?

Morie

This is a very tall order. I can list some of the basic detectors, but a complete list has not been worked out yet.

For the sake of our discussion, let us first distinguish between sensors and detectors. A sensor detects the most basic element of information – information from a pixel. For example, a cone cell in the retina senses a point of light. A pressure-sensing cell in the skin senses the pressure on a small skin element. On the other hand, a detector detects information provided by several sensors or by several nodes. For example, a line detector detects cues provides by many point sensors. A danger detector that detects alarming sounds is fed by signals from many auditory sensors.

Our perception is based on signals from various sensors including the six senses, sensors of state of our muscles and joints, sensors of molecules that are involved in metabolism, and sensors that sense our feelings. Detectors detect cues relayed by those sensors and by other firing nodes in the brain.

There are detectors that operate is a certain modality, such as visual detectors or auditory detectors. Other detectors operate across modalities. Let me start with basic detectors that operate almost in all modalities. Some of these detectors are so prevalent, that we tend sometimes to take them for granted.

Becky

Do the same neural circuitries underlie all the detectors that do the same task?

Morie

Not necessarily. However, at the level of our description of the system, we treat detectors as parts of nodes, without addressing their internal mechanisms. Different detectors may rely on different circuitries and mechanisms. Such details are outside the scope of our discussions.

SAMENESS DETECTORS

Morie

When we look at a clear sky in a bright day, millions of sensors in our eyes sense the color blue. A detector tells the rest of the brain that all those sensory pixels are receiving the same stimulus. Based on that information, we perceive the sky as one blue item.

Sameness detectors aid in the separation of an entire scene to items, each of which is made up of the same parts. Detectors detect sameness of colors, sameness of texture, sameness of motion, etc. A hawk gliding in the sky would spot her camouflaged prey when it moves. The hawk's sameness detectors grouped together similar visual signals that were triggered by the motion of the prey.

All of our senses feed sameness detectors and the information provided by sameness detectors affect our overall perception. We can tell when our feet are too cold and the rest of our body is fine, thanks to the few sameness detectors that monitor stimuli from multitudes of skin pixels.

Sameness detectors operate also on information that they receive from intermediate neurons. They can tell when two items are exemplars of the same class, as in "this is an apple and this is another apple".

Sameness detectors can indicate when different patterns of neurons that have fired at different times are instances of the same item. For example, when a ball is passed to us, the patterns of visual neurons that it stimulates keep changing continuously. Certain sameness detectors group the ball's pixels at each instant of time, and other sameness detectors tell us that all those items are instances of the same item.

Sameness detectors are also involved in distinguishing between novel stimuli and familiar ones. The novel stimuli originate from the sensory units, whereas the signals of the familiar counterparts originate from the memory. Sameness detectors participate in identifying familiar items in the incoming information stream.

OPPOSITES

Morie

In order to survive, an organism must distinguish between three kinds of stimuli: 'good', 'bad', and 'benign'. When 'good' stimuli are detected, the organism should behave so that those stimuli continue. When 'bad' stimuli are detected, the organism must adjust its behavior to ward them off. Benign stimuli could be ignored. These distinctions are accomplished by two basic detectors – 'good' and 'bad', which relay their signals to appropriate activity-plans.

Using innate learning mechanisms, each of the two basic detectors 'good' and 'bad' establishes its own class and adds exemplars to it. The first exemplars are basic sensations such as hunger, being fed when hungry, excessive heat, comfort, wetness, etc. With time, the exemplar-lists are expanded to include also activities and activity-plans that were associated with good or bad feelings.

Exemplars of 'good' and 'bad' may be exemplars of other classes as well. That creates sub-classes of the main 'good' and 'bad' classes e.g. 'good sensations', 'bad responses' etc.

When two items, an exemplar of 'good' and an exemplar of 'bad', belong to the same class, they constitute a pair of opposites. For example, the pair 'tasty food' and 'disgusting food'; 'tasty food' is an exemplar of 'good', 'disgusting food' is an exemplar of 'bad', and both are exemplars of the same class, 'food'.

Each exemplar of the class 'opposites' is an item that has two parts. For example, 'black' and 'white' are two parts of the node 'black and white', which is an exemplar of the class 'opposites'. (The class 'opposites' itself is an exemplar of the class 'mutually exclusive'.)

Initially, members of the class 'opposites', through their associations with 'good' and 'bad', have emotional connotations. With time, the class 'opposites' expands to include pairs of emotionless concepts such as 'warm and cool', or 'high and low'. The members of an opposite pair must be exemplars of a common class.

Very early in our infancy, we learn the association between 'good' and 'yes' and between 'bad' and 'no'. Initially, 'yes' and 'no' have emotional associations: "yes, play with mommy", "no! Don't play with the electric cord". With time, 'yes' and 'no' become associated with emotionless opposites and serve as the archetypical opposites.

CONFLICT DETECTORS

Morie

Sentences are made of words and implied instructions on how to fuse them into one inclusive concept. The words, which by themselves are concepts, are represented in the brain by nodes. The brain translates sentences into nodal representations, and incorporates them in its information structures.

For example, consider how the sentence "Tom's face is beautiful and tanned" could become a part of the brain's information structure (Figure 11). The new concept 'Tom' and the existing concept 'face' are fused to form the new concept 'Tom's face'. Then, the existing concepts 'beautiful face' and 'tanned face'

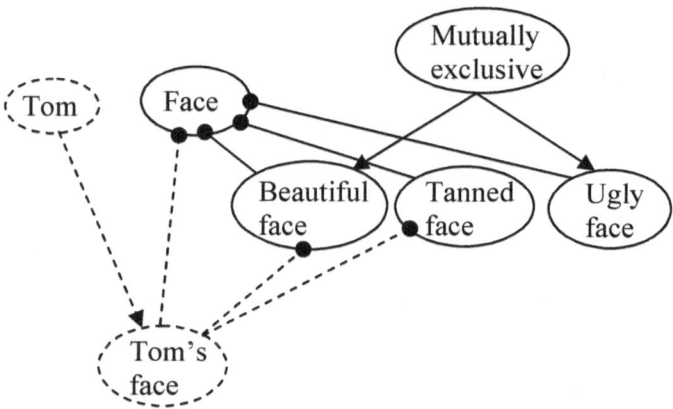

Figure 11 Incorporating a sentence with the existing information structure. Dashed lines indicate new information, solid lines – existing information.

are fused with 'Tom's face', making the entire sentence an integral part of the brain's information structure.

Using the same mechanisms, the nodal representation of the sentence 'Tom's face is beautiful and ugly' is shown in Figure 11A.

The difference between the two sentences is that the second

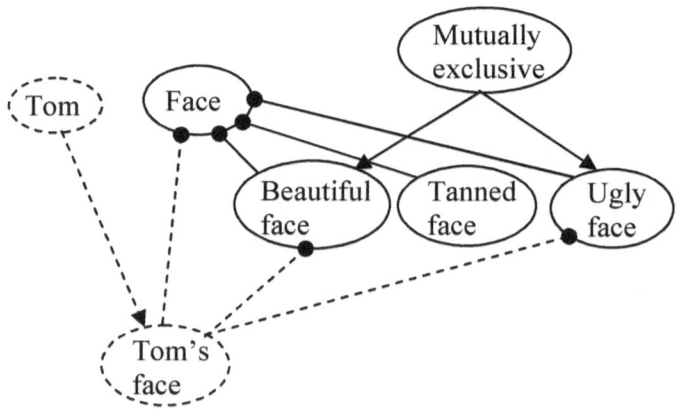

Figure 11A An information structure with a conflict

has a conflict. The activity-plans that the brain uses to translate sentences into nodal representations allow for the creation of concepts with internal conflicts. Conflicts are the result of various situations. A common situation is assigning an item as an exemplar to two opposite classes. In Figure 11A, 'Tom's face' became an exemplar of the opposite classes 'beautiful face' and 'ugly face'. The mechanisms that the brain uses for representing a sentence do not check if the sentence has conflicts.

The brain may employ separate conflict detectors to check for conflicts in its information structures. Those detectors notify other activity-plans about their findings, and corrective measures may ensue.

Claire

That reminds me the opening of Dickens's "A Tale of Two Cities": "It was the best of times, it was the worse of times..."

Morie

Oh, yes, unforgettable. You could feel the conflict-detectors beating at the end of each contradiction: "It was the best of times, it was the worst of times, it was the age of wisdom, it was the age of foolishness, it was the epoch of belief, it was the epoch of incredulity, it was the season of Light, it was the season of Darkness, it was the spring of hope, it was the winter of despair, we had everything before us, we had nothing before us, we were all going direct to Heaven, we were all going direct the other way—". All those conflicts are created by assigning an item as an exemplar to two opposite classes.

Sol

Are all conflicts created by assigning an item as an exemplar to two opposite classes?

Morie

No. Opposite classes are a special case of a wider group of classes —mutually exclusive classes. Classes are mutually exclusive if they cannot have a common exemplar. For example, 'red ball', 'green ball', and 'yellow ball' are mutually exclusive classes, because a ball cannot belong to more than one of them. Assigning an item as an exemplar to two or more mutually exclusive classes creates a conflict.

In addition to direct conflicts, there are also implied conflicts. An implied conflict is created when an item is incorporated into the network, and a conflict emerges due to nodes that were not involved directly with the incorporation.

For example, Sarah and Chris have a new baby, Jonathan. This information is recorded in the brain of their friend, Theodore, as 'Jonathan is the son of Sarah' and 'Jonathan is the son of Chris'. As Theodore goes on with his own business, his brain keeps thinking in the background, using information recorded in his information structures. He knows that Sarah is Jewish and Chris is Christian. Cues in his brain would classify as 'Jewish' a baby whose mother is Jewish and as 'Christian' a baby whose father is Christian. A conflict detector is triggered in Theodore's brain. Jonathan is an implied exemplar of mutually exclusive classes 'Jewish' and 'Christian'. Theodore is baffled. He would have to rethink this part of his information structure.

Claire

It is possible that Theodore might have never become aware of that conflict.

Morie

Right. Conflicts in our information structures may go undetected forever.

Bob

How does the brain resolve conflicts, once it detects them?

Morie

The brain may leave detected conflicts as they are, or it may modify the information structure in order to resolve them. It may modify cues of existing classes, so that the problematic item would not belong to more than one mutually exclusive class. For example, Theodore may define new cues for the religious affiliation of newborns. It may define religion of newborns as the way they are raised by their parents, or as undefined until they grow up and decide for themselves.

Becky

Those kinds of conflicts are manmade. They are created because our brain can incorporate verbal information. Are there naturally created conflicts?

Morie

Consider the information in a mouse's brain. A mouse knows that it has to get to a piece of cheese, in order to survive. It also knows that it should not go outside of its hole if the cat is around. Should it get out to the cheese when the cat is in the other room?

Bob

Such a conflict may drive the mouse crazy.

Morie

That's what recorded conflicts do sometimes, when they clash with reality.

DATA PROCESSING DETECTORS

Morie

Since information is stored by the brain in associative structures, information-retrieval requests may result in a variety of outcomes. An outcome may consist of several possible nodes, of one node, or none.

Any retrieved node may answer the request completely, in part, or not at all. In dynamic retrieval processes, the retrieved result may be in conflict with the original request.

Basic detectors have to assess every outcome according to those possibilities and forward their findings to the appropriate activity-plans that handle the information. The retrieval-detectors may have to utilize in their evaluations conflict-detectors.

TEMPORAL DETECTORS

Morie

The sensation of time is one of our basic sensations. A number of basic detectors contribute to our perceptions of time.

Three basic time detectors detect temporal features in events of all modalities: the 'begin', the 'end' and the 'going-on' detectors. As their names imply, those detectors, respectively, fire when a node in any modality starts to fire, stops firing, or keeps-on firing. The temporal features that they detect are used by other temporal detectors to further characterize the events.

The temporal-order detectors, 'before', 'after', and 'at the same time' detect the temporal relationship between two events. They depend on information provided by the first three detectors.

By combining the temporal-order detectors with the 'now-detector', the detectors of past events, current events, and future events are constructed. For example, a past event is an event that ends before 'now'.

If we are told that Abe was born before Ike, and Ike was born before Jack, we may conclude that Abe was born before Jack. In general, if "A" happened before "B", and "B" happened before "C", then we conclude that "A" happened before "C". This conclusion is derived by a compound temporal detector that relies on all the temporal detectors mentioned so far.

Whereas all those temporal detectors provide discrete information, time duration detectors provides graded information about the length of time intervals e.g. very long, long, short, very short time. The determinations of these detectors are subjective, and they depend on the situation. For example, three minutes may seem a short time for a child in a fun ride in an amusement park, and a long time on the receiving end of the dentist's drill.

Sol

What are the relationships between the detectors 'before' and 'after'?

Morie

We know that for any two given events, "A" and "B", '"A" happened before "B"' is the opposite of '"A" happened after "B"' and is the same as '"B" happened after "A"'. All those relationships are encoded by the brain's networks.

ORDER DETECTORS AND MATHEMATICS

Morie

The temporal detectors that we have just described are the basis for our perception of ordered events. In addition to operating on incoming signals from the outside, the detectors can operate also on retrieved information and simulated events, and extract details about their internal order. For example, we "get the picture" when we are told, "plants existed before animals", or "Adam was created before Eve", even though we have not witnessed it.

The brain can create simulations in which events occur in any arbitrary order. Those simulations could be analyzed by other detectors, and conclusions may be drawn from such analyses. These abilities of the brain underlie our understanding of counting, and everything else that depends on it.

Natural numbers are abstract concepts that are arranged in a sequence. We learn that 1 comes before 2, which comes before 3, and so on. That means that we have assigned temporal properties to the sequence of numbers. When we count apples, we playback the sequence of numbers and assign them, one by one, to the apples. We thus assign temporal properties to activities that we do with the apples as we count them. When we say that 5 apples are more than 3 apples, we base it on the information recorded in our brain, that when we recite the sequence of numbers, 5 appears after 3.

The brain can create simulations that combine counting with other operations. That is how the concept of counting is expanded, and general properties of numbers are derived. Abstract mathematics has its roots in our basic perception of numbers, which is directly related to our basic temporal detectors.

Counting is at the foundation of science. Hence, all our mathematical and scientific perceptions hinge on the brain's temporal detectors and simulators.

STATUS DETECTORS

Morie

The brain has basic detectors that are used to detect the status of body parts or the progress of internal processes. We will call them 'status' detectors. Those detectors participate in detecting concepts that, on the surface, have nothing to do with the status of the body. Let us see an example of how status detectors combine to detect an "unrelated" concept.

Basic detectors detect the state of whole muscles; whether and to what extent the muscle is relaxed or contracted. Such detectors receive signals from sensors that sense muscle cells. One of those status detectors detects the contraction of the muscles at the back of the neck. We may call it the 'nape contraction' detector.

In order to see clearly, eye muscles adjust the lenses in our eyes so that a sharp image of what we look at is formed on the retina. Once a sharp image is formed, those eye muscles stop their adjustment. A detector must be detecting when a sharp image is achieved. Let us call this status detector the 'in focus' detector.

The 'above' detector (as in 'A is above B') is one of the many compound detectors that detect features of external objects, based on outputs of internal status detectors. Several mechanisms may detect when object A is above object B. Consider one of them that detects an airplane above the ground.

When we detect that an airplane is above the ground, the following two sub-detectors are in action: First, we focus our eyes on the flying airplane. The sub-detectors and their triggers that participate in this state form the following combinations: 'air plane is in focus' AND 'nape muscles are contracted'. (This is the state of those muscles when we look up and see the airplane clearly.) Second, we focus our eyes on the ground. The involved sub-detectors and their triggers are: 'the ground is in focus' AND 'nape muscles are relaxed'. We say that one object is above a second object when the first object is in focus and the neck muscles are contracted, and the second object is in focus and the neck muscles are relaxed.

'Above' is an example of a concept relating to the external world that is defined by detectors that detect the state of internal body parts and processes. 'Above' is a spatial relationship between two external objects. It is expressed as a combination of fundamental detectors that detect internal information; the state of the muscles of the nape and signals from the focusing mechanisms of the eyes' lenses.

OTHER DETECTORS

There are many more basic-detectors. Some of them are modality specific. For example, the 'inside' and 'outside' detectors detect those features from streams of visual information. We will describe few additional detectors in our forthcoming discussions, but for the most part, a comprehensive list of all the basic detectors is still waiting to be compiled.

THE LIMITS OF HUMAN PERCEPTION

Sol

The picture that emerges from our discussions is that the innate brain has some information in it. In addition, it has machinery for processing and acquiring new information and for constructing new processing tools. The basic detectors and activators process incoming stimuli and create new detectors, activators, and records of our experiences. Since all those elements build our perception, it seems that our perception might be restricted. If there are events or regularities in the world that could not be detected by our detectors, we would be unable to perceive them.

Becky

What exactly do you mean? Our eyes cannot detect x-rays, but we can still perceive them and use them. We cannot see atoms, but we can perceive them.

Sol

X-rays are just one of many artificial extensions of our senses. Intermediary devices translate their signals into signals that could be sensed by our senses and then processed by our detectors.

Atoms are too small to be seen, even by a microscope, but we can perceive them through models. At the beginning of the twentieth century, scientists developed models of the atom. According to those models, the atom is like a small solar system, in which electrons are orbiting around the nucleus, similarly to the planets that orbit around the sun. While predictions about the motions of planets turned out to be very accurate, predictions about atomic processes, which relied on the models of the atom, conflicted with observations. So, new models that are based on counter-intuitive assumptions were developed. For example, the models assume that an electron behaves as a wave and as a particle; something that we do not encounter around us. In general, the world at the atomic and sub-atomic scales has to be perceived differently from our macroscopic world. However, we perceive both worlds by concepts that were developed in our brain by the same basic detectors.

My point is that both our innate detectors and all our acquired concepts, which are built by them, might be unable to detect some of the basic laws of nature. Therefore, we might be unable to perceive those aspects of the world.

Becky

I think that I begin to see your point. We can perceive colors due to the neural networks that connect three kinds of light-sensitive cells in our eyes. Imagine that humans, like many other animals, did not have those cells and those circuits. Then we would not have color vision. We would see only shades of grey. Would we be able then to perceive colors the way we perceive them now?

Morie

That is a good example. Let me add to the dilemma. Certain detectors are so central to our existence, that we have come to believe that they represent laws of nature. One example is the cause detector. This compound-detector identifies the cause of events. Since such information is crucial for planning our activities, we assume that every event in the world must have a cause. Each scientific discipline tries to uncover what are the causes of the phenomena on which it focuses.

Is it a universal truth that every phenomenon, without exception, has a cause? Or is there an exception, or exceptions, and there are phenomena that happen "out of the blue", without any cause. Along the same line, our experience tells us that everything is preceded by other things. Are there primary phenomena that nothing preceded them? If there are such phenomena, with which detectors can we perceive them?

Claire

Do you mean God? Nothing preceded God. God is the first cause.

Morie

This is one example.

Bob

I do not see your point. I believe that I perceive God, even without detectors that can identify phenomena that have no precedent or no cause.

Morie

There is a saying that Man created God in his own image. We can perceive God, because our brain can assemble arbitrary parts that we provide it, and define an item out of them. It can assign that item to classes, and it will inherit additional properties from those classes. The brain can incorporate this item in its information structure. As such, the item can be part of our thoughts, participate in mental simulations, and invoke emotions. By invoking emotions, it may affect how new information is incorporated in our mind. What we believe in, and what we reject. The brain can modify and fine-tune its information structures, in order to deal with conflicts that might be created by the incorporation of that item. It can define new classes that deal with super-natural entities, so that conflicts with natural entities are resolved. In short, the basic information-handling-mechanisms of the brain can do all of that, thus enabling us to perceive God and other super-natural entities.

Bob

Does not it prove that the brain is so powerful, that it can perceive all the natural and super-natural phenomena?

Morie

While we can perceive some aspects of super-natural entities, those basic mechanisms might be insufficient for perceiving everything that exists in the universe.

In general, our perceptions are created by detectors in our brain, which are designed by our genes and are affected by our experiences. There is always the possibility that such detectors and their derivatives would be unable to create perceptions of everything that happens in the universe.

UNDERSTANDING

BLOOM'S TAXONOMY

Claire

Perception goes hand in hand with understanding. How understanding is realized in our model?

Morie

Educators need to understand understanding. In 1954, a group of college educators, after meeting and discussing that issue, published a taxonomy that contains six different types of understanding, organized according to their inter-dependencies. The six groups are *knowledge, comprehension, application, analysis, synthesis and evaluation*. That list has since become widely known as Bloom's Taxonomy, named after the meeting's organizer and the report's editor. The report defines learning objectives for each category of understanding and describes ways of enhancing and evaluating the effectiveness of the learning process for each category.

At the core of the issue is the way that the brain stores and uses information. Anything that we learn, be it on our own or by being taught, is only a partial representation of the whole picture. When we use information that we have learned about a specific subject, we usually supplement it with general information, utilizing certain brain processes. The outcome demonstrates how well we understand the specific subject. If the outcome conforms to reality or to similar outcomes derived by others, we say that we understood the matter. Otherwise, we did not understand, or even worse, we misunderstood it. Let's illustrate with an example those six levels of understanding, and their underlying mental processes.

EXAMPLE

Emma is the mother of Ben, a third grader. Ben joined a soccer team, and Emma car-pools him. Emma does not know anything about soccer, and she decides to learn the subject 'on the fly', by just watching the games. After the first game, she gets the drift of it. There are two teams. Players pass the ball to each other by kicking it. The objective is to kick the ball into your opponent's goal. When that happens, your team scores a point. Another objective is to prevent the other team from scoring. Touching the ball with the hands is not allowed, except by the goalie, or in order to throw it back into the field, after it has crossed the sidelines.

Knowledge: Knowledge is the lowest category in Bloom's Taxonomy. It is the ability to recall learned information.

In Emma's case, knowledge is familiarity with the basic elements of the situation and the relationships between them: players are running and kicking the balls, trying to score a goal, and trying to prevent the other team from scoring. Emma can tell when a goal is scored, when a penalty is called for touching the ball, etc.

Nodes have been recruited and new connections have been established in Emma's brain in order to represent the learned information. Nodes represent the new concepts, such as soccer-field, what players are doing and what they should not do, etc. These nodes, which act as detectors, are activated when their corresponding events happen on the field.

Comprehension: "Comprehension", which is the second level, is the ability to give meaning to information categorized as "knowledge". For example, whenever her team scores a goal, Emma is excited. She is able to correctly assign her emotions to the particular situation on the field.

Micropsychology

Claire

Assume that before the game someone has told Emma to jump with joy whenever her team scores. Would that piece of information be considered "knowledge" or "comprehension"?

Morie

It would be considered "knowledge", because it would be part of the information that is recorded explicitly in her brain. In "comprehension", the meaning of a recorded piece of information is determined after combining it with prior independent knowledge. In Emma's case, she is already familiar with the concept of winning, how scores are kept in sports games, and she likes winning. She combines all that prior knowledge with the new knowledge about soccer and the particular information that her team has just scored a goal, and gets the right meaning of scoring.

Sol

How this is done by Emma's nodes?

Morie

In general, the process is classification. In Emma's brain, a hierarchy of classes underlies that response. When an exemplar is activated, it activates its classes, which activate their over-classes. 'We score a goal in soccer' is an exemplar of 'we score in a game', which is an exemplar of the top class of the hierarchy; 'a joyous event'. Whenever her team scores, the event is classified as an exemplar at the bottom of that hierarchy. Signals propagate up the hierarchy and eventually activate the top node 'a joyous event'.

Application: Application is the third level in Bloom's Taxonomy. In application, learned information is adapted to other circumstances.

In the next weekend, Emma drives a group of kids to the beach. There, she organizes an impromptu soccer game. Under Emma's instructions, the kids mark in the sand the boundaries of a soccer field, pile sand mounds to indicate the goal posts, and set up two four-player teams. The kids play the game without a referee, guided by their natural sense of honesty. This impromptu game is an example of the third category of Bloom's Taxonomy; "application". Known information has been adapted to different circumstances.

Here also, the underlying basic nodal mechanisms involved class-exemplar operations. In Emma's brain, some nodes of the original information-structure about soccer were replaced by other exemplars of the same class. For example, each team now has four instead of eleven players. 'Eleven players' and its replacement, 'four players', are two exemplars of the class 'number of players'. Similarly, 'Beach' replaced 'grass'. Both are exemplars of the class 'flat playing area', and so on.

Analysis: Analysis is the fourth level in Bloom's Taxonomy. In analysis, thoughts are focused on parts of the original information structure. Implicit relationships between those parts and previous concepts are uncovered and used in the particular circumstance, or become an explicit part of the information structure, which may be used in the future.

Emma watches the kids play. She notices that Ben usually gets the ball close to his own goal, but very rarely he can dribble it to a kicking distance from the opponent's goal. If he dribbles for too long, he gets tired and looses the ball to fresher opponents that tackle him. If he passes the ball to other players of his team, the ball is intercepted quite often. She concludes that Ben has difficulties with his dribbling and passing. She also notices that

Garry is a slow player, but he has a very strong kick. This type of information processing is categorized as "Analysis".

In Emma's brain, connections are developed between the analyzed nodes and nodes of other origins. For example, a new concept, represented by a new item node, is formed: 'Ben gets too tired after dribbling too long'. The parts nodes of this item include the nodes 'Ben', 'dribbling', 'too long', and 'tired'. Some of those parts have been learned by watching soccer, while others have been learned in unrelated events. All those nodes and the connections between them are formed by the basic brain mechanisms that assemble individual parts into new items, and give meanings to items by making them exemplars of classes.

Synthesis: Synthesis is the next level. In synthesis, information from lower categories is combined to create new information entities. Synthesis usually relies on analysis, because information that was generated by analysis is used in assembling the new entities.

In the drive home, as the kids are napping in the back, Emma is thinking. She wants Ben to be a good player. She imagines Ben as a star soccer striker, playing on the varsity team of his high school. Scouts discover him and he gets a scholarship to an Ivy League university. In the championship game (nationally televised), everyone in the stadium points at her and whispers: "this is Ben's mother". As the mental simulation wanes off, she starts thinking about realizing her dream. Ben has to become a star in his little-league team. She plays in her mind all the original and deducted information about soccer. How can Ben score more goals? And then she synthesizes a solution: Ben should stand next to the opponent's goal, waiting. Garry, with his strong kick, will pass the ball down the field to Ben, in a "hail Mary" kick. Then, Ben will send it straight into the opponent's goal. He won't have to dribble too much or to pass it to other players. Her problems are solved!

In Emma's brain, a node was recruited to represent her new game strategy. That node represents a sequence of events, each of which is an element taken from the original information structure: 'Ben stands near the opponent's goal', 'Garry passes the ball to Ben', 'Ben scores'. Each of those nodes has connections to other nodes, which were derived at the analysis level e.g. 'Ben gets the ball when he is not tired'.

Emma invites Garry home, and over a sumptuous bowl of ice cream she shares with him and Ben her brilliant plan. Garry likes it, because he does not like to run and he is promised that the 'assist' of each goal would be credited to him.

Next Sunday they all go to the match. Ben plants himself next to the opponent's goal, ignoring the pleas of his coach to get backfield and help his team in the skirmishes. The big moment comes. Garry gets the ball, and with an explosive kick sends it smack to Ben. Ben turns around, as a whistle is heard, and nails the ball into the net. Emma jumps up and down and screams with all her lungs: Goooooooaaaaalllll! as she has seen on TV, in "South-American Soccer". (In Bloom's Taxonomy, this part of her behavior would be categorized as "application"). But she is the only one jubilating.

Bob

Uh, oh. It is not a good sign when you are the only one laughing at your own joke, clapping your hands in a classical music concert, at what you consider to be the end of the piece played, or jumping up and down at a sports event.

Morie

Right. The referee calls offside! No goal! Emma goes berserk. She screams at the referee, "What's you problem? Don't you see? This is a goal!!!" She is calmed down by people around her, who mutter that the referee was right, and that Ben has committed an unquestionable offside. Her game plan, which could be categorized as "synthesis" in Bloom's Taxonomy, is categorized by the other parents as "misunderstanding".

Evaluation: The sixth level in Bloom's Taxonomy is evaluation. The value and merits of the learned information, as acquired explicitly and as enhanced by the other processes, is evaluated. In evaluation, the benefits of the learned and other recorded information are compared with each other.

At home, Emma thinks over her entire involvement with Ben's soccer. She weighs her decisions and actions and how they match her expectations, beliefs and moral values. She decides to learn the rules of soccer in a systematic way from reliable sources, and to behave appropriately in Ben's soccer games.

In evaluation, the nodes that represent the learned information are activated and their effects on selected detectors are observed. For example, when Emma simulates in her mind her reactions to Ben's offside kick, the node that represents 'embarrassment' is activated. That associated negative value is the first step in her cognitive processes that would lead her to her conclusions.

Claire

The taxonomy that you have just described deals mainly with merging cognitive operations with information that can be traced back to sensory inputs. Are there taxonomies that deal with other aspects of brain's function, such as emotions and motor activities?

Morie

Bloom and his associates introduced a second taxonomy for

affective learning, which deals with the organization of learned emotional skills. Other researchers introduced taxonomies of learned motor skills, and also various modifications and alternatives to the original Bloom's Taxonomy. Those taxonomies are widely used by educators, because they organize target skills by the order of their complexity and interdependencies. That facilitates the design and assessment of learning plans.

The same basic brain mechanisms underlie the learning and the utilization of cognitive, affective and motor knowledge. All learned information is organized in classes-exemplars and items-parts hierarchies and is manipulated by the same basic mechanisms that operate within those structures, regardless of the meaning of the stored information.

BASIC NODAL PROCESSES AND THE TAXONOMY

ITEMIZATION AND CLASSIFICATION

Sol

Bloom's Taxonomy ranks the different categories of understanding by their complexity, e.g. 'analysis' is higher than 'knowledge'. Yet, the same underlying brain mechanisms are involved with all the different categories. Is there a relationship between the complexities of the understanding processes, as implied by their ranking in the Taxonomy, and the underlying brain mechanisms?

Morie

The same nodal-level mechanisms underlie all the levels of Bloom's Taxonomy. However, different aspects and different combinations of the nodal mechanisms are applied in the different levels.

Itemization and classification are two basic brain mechanisms that are involved with all the categories of the Taxonomy. In itemization, separate information entities are grouped together to form an item, represented by a node. The constituting entities become part nodes of that item node. Itemization reduces the number of entities that the brain has to handle in order to understand a situation. For example, when Emma watches a game, she has to follow what players are doing, but she does not have to pay attention to how their muscles are contracting.

In classification, a node becomes an exemplar of a class. It thus inherits the common properties of that class. Classification makes it possible to use information that has not been spelled out explicitly. For example, no one has told Emma to be jubilant when her team scores. Nonetheless, Emma responded with joy when her team scored a goal. This happened because her brain classified the goal as an exemplar of the pre-existing class 'my team scores in a game', which triggered a joyful response. The ability of the brain to classify incoming information into existing classes is one of the most important factors that determine the understanding power of the individual.

DETECTORS IN THE TAXONOMY

Morie

Classification is carried out by specialized detectors. Each class has its detector that receives streams of signals that come from the outside or from certain regions of the brain. A detector is tuned to a specific pattern of signals. When such a pattern appears in the stream, the detector is triggered and identifies the nodes that sent the triggering pattern. For example, when we read a book and a fly lands on it, we detect it immediately. A motion detector in our vision system detected the moving object within our field of view, and identified the pixels in which the motion occurred.

The outcomes of the classifications are merged with the information stream. For example, when Emma watches the kids play, pixels in her field of view are first itemized into separate moving objects. Then, items are classified by the detectors in her brain into several classes such as 'child', 'player of my team' 'player of the opponent', 'goalie' and so on. That information is represented explicitly by nodes and by the connections between them. For example, the node 'Garry' is an exemplar of the classes 'child' and 'player of my team'.

Detectors may classify items according to a wide variety of criteria. Some detectors classify items according to their relationships with other items. For example, 'Ben tackles John' is an exemplar of the class 'player tackles another player'. The classification is done by the detector of that class, which detects when such an event happens in a game.

RETRIEVAL IN THE TAXONOMY

Morie

Let's see now how itemization and classification are implemented at the different levels of the Taxonomy.

In operations at the *knowledge* level, retrieval requests address only information that is expressed explicitly in the memory. For example, Emma's comments: "Ben was tackled by a player of the other team", or "the goalie caught the ball that Ben kicked" are recalls of information that was recorded explicitly in her brain.

In operations at the *comprehension* level, items from the original information structure activate class nodes that are not part of the original information structure. For example, the underlying nodal process that caused Emma to be elated when the goal was scored was a sequence in which class-nodes were activated by their exemplars. The actual event was classified as scoring, which was classified as a joyous event, which triggered the response. The latter two activated nodes are not part of the soccer-information-structure, as learned by Emma.

In operations at the *application* level, an item from the learned information body is replaced by another exemplar of the same class. In order for this to happen, the item has to be classified, which is an operation at the *comprehension* level, and then another exemplar of the identified class has to replace the original item. That was the underlying nodal process when the original grass field was replaced by sandy beach; both are exemplars of the class 'flat playing area'.

In operations at the *analysis* level, a retrieval request retrieves information that is associated with elements of the original information structure. This is somewhat similar to *comprehension*. The main difference between the two is that the retrieval requests in *analysis* are more complex than those in *comprehension*. For example, a retrieval request in *analysis* may be: "what causes Ben to lose the ball?" This is a compound retrieval request that can be resolved into basic requests.

Sol

How this particular information could be recorded in Emma's information structure, and how the retrieval request could be resolved into basic requests?

Morie

The cause detector plays a central role in this process. It detects when one event, e.g. event A, is the cause of a second event, e.g. event B. When that detector is triggered, it creates a record by recruiting nodes and establishing connections between them, so that they represent the information that event A was the cause and event B was its effect.

Figure 12 illustrates an information structure representing that Ben's fatigue caused him to lose the ball. Solid lines indicate the network of class nodes to which the new information is assigned as exemplars. That establishes the meaning of the new information. Dashed lines indicate the nodes and the connections that were recruited by the cause detector to record the information item: 'Ben's fatigue caused him to lose the ball'. Round arrows indicate exemplar-class connections, and pointed arrows indicate item-part connections.

The question "what caused Ben to lose the ball?" is expresses as the following retrieval request: Retrieve [exemplar of 'Cause' AND [part of '[exemplar of 'Cause and Effect'] AND [item of (['Ben lose ball' AND [exemplar of 'Effect'])]]]].

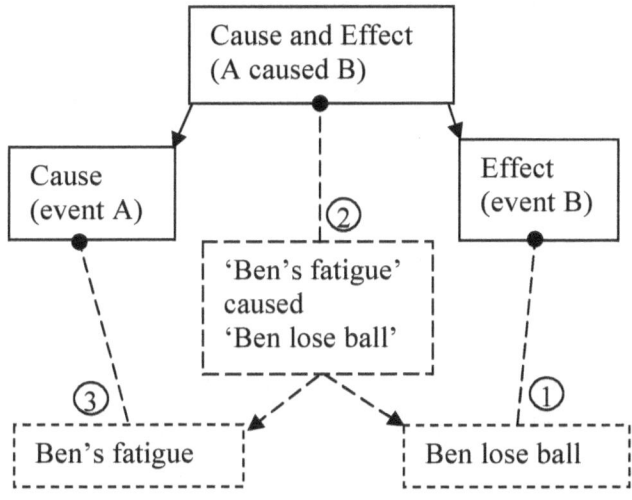

Figure 12 Incorporating a cause-effect pair with the existing information structure. Solid lines indicate existing information, dashed lines, new information.

To follow the retrieval request, we start from its end: (['Ben lose ball' AND [exemplar of 'Effect']). This will activate a node (marked as (1) in the figure) that represents Ben losing the ball as the result of some unspecified cause. Now, that this node is active, the request '[exemplar of 'Cause and Effect'] AND [item of (['Ben lose ball' AND [exemplar of 'Effect'])] would cause the firing of 'Ben's fatigue caused Ben to lose ball'(marked as (2) in the figure. With this node firing, the rest of the retrieval request would retrieve the answer 'Ben's fatigue'.

Claire

In similar ways, *synthesis* and *evaluation* operations could be resolved into basic retrievals, but their combinations would be more complex than those of the lower categories of the Taxonomy.

Morie

Right. In *synthesis* the final outcome is a new information entity, which is composed of original information and information that was derived by lower categories operations. In *evaluation* the retrieval requests may be based on compound cues from both the original and other information structures.

Bob

What came out of Emma and Ben?

Morie

Emma took a course and became a soccer coach. Her players start each practice with grueling twenty minutes of running and sprinting, which have improved significantly their stamina. Ben is very good at video-soccer.

INTELLIGENCE

Claire

We have seen how the mechanisms by which information is

perceived, recorded, and accessed by the brain affect understanding and behavior. We all employ those same mechanisms. Why are some of us more intelligent than others?

Morie

Our intelligence depends on the effectiveness of those basic mechanisms and on the thinking apparatus that deploys them.

While every brain uses detectors that detect cues in information streams, the sensitivity of the detectors may vary from one individual to another and affect the overall intelligence of a person. Differences in synaptic weights affect the intensity of the signals needed to trigger a detector, thus affecting the sensitivity of the detector.

As the brain processes and records information, it needs to form and modify synapses. Although all those processes rely on the same basic mechanisms, the efficiency of the mechanisms may vary between individuals.

Due to those individual variations, the amount of recorded information and its details vary between individuals. The same event may not be remembered by some individuals, may be remembered vaguely by some, and may be remembered with fine details by others. It all depends on how effectively the incoming information stream has been itemized and how the items were classified. Both itemization and classification require the modification of existing ties and the recruitment of new ones. So, the detectors in the brains of intelligent people are more effective in itemizing and classifying the incoming information stream. The synaptic modification mechanisms in the brains of intelligent people are more effective in recording the information that was detected by the detectors.

Once the information is stored, the brain of intelligent people retrieves it easier, by using the same basic retrieval mechanisms.

Retrieval in the brain is done by association. Given information, which is expressed by active nodes, retrieves associated information by activating quiet nodes. The activated nodes are related to the given nodes through direct and indirect item-part and exemplar-class relationships. However, only some of those associated quiet nodes are activated immediately by the given nodes. Some of the quiet nodes are activated only after longer retrieval procedures, e.g. by modulation. Some associated nodes may not be retrieved at all.

For example, imagine Al walking down the street and meeting Alice, whom he has not seen for a while. He is happy to see her, and they hug each other. The eye pixels that are triggered by seeing Alice activate the item-node 'Alice' in Al's brain. The firing node 'Alice' then activates the class-node 'someone that I like', which makes Al happy.

'Someone that I like' was activated due to its association with 'Alice'. Other nodes that are associated with 'Alice', such as 'Alice's cat' were not activated, due to their weak or indirect connections with 'Alice'.

Al's action of hugging Alice was the result of a conscious thinking process. It took effect because the associated node 'someone that I like' was activated, and could participate in a thinking process 'what should I do now?' On the other hand, Al did not ask Alice about her cat; not because he did not care about it; frankly, the 'cat' node was not triggered, so it did not take part in Al's thinking processes.

INFORMATION HALO

Morie

An event activates nodes that represent it. In addition, secondary nodes that are associated with the original ones are activated.

Those secondary nodes may be called the halo of the event. (e.g. 'Alice' in the last example was the event, and 'someone that I like' was part of the halo.) The halo contains also associated nodes that are quiet, but they have been partially stimulated by the original ones.

The first response to an event, which may be the only response, is based on nodes from the halo, because they are associated most closely with the ongoing event. Other nodes would be considered, only if nodes from the halo did not provide an appropriate response. The latter would be a lengthier process, requiring activity-plans that are more complicated, and it may not yield any results.

Intelligence is determined by responses to given situations. Therefore, the width of the halo is one of the factors that affect intelligence. It makes potentially relevant information more accessible to thinking activity-plans. If a halo is too narrow, not enough associated events would be readily available for consideration. If it is too wide, too many associations will pop up. They will have to be sorted out, complicating the response processes.

The width of a halo depends on the synaptic weights between the nodes that represent the main event and their associated nodes. In any individual brain, the width of the halo may depend on the type of the information considered. For example, people adapt at mathematics have a wider halo surrounding the representations of mathematical concepts in their brains. People that are mathematically challenged have much narrower halos around nodes that represent mathematical concepts. That is why "no bell rings in their head" when they are asked a math question.

Sol

What is the role of the quiet nodes in a halo?

Morie

The halo may include also primed nodes. Primed nodes receive signals from the main event, but not enough to fire them. Because they get closer to their firing thresholds, thinking processes would retrieve them more easily.

Nodes that originally have not been included in the halo may still be accessed by deliberate retrieval processes. If the retrieval relies on modulation, the modulation rate of the more intelligent brain is fast and has a high resolution. It activates the right number of nodes in the shortest time.

Bob

Because of their wider halo, does it mean that people who know mathematics are smarter than the rest of us?

Morie

Absolutely not. The size of the halo is not something fixed all over the brain. It may vary from one brain region to another. The halo of poets in brain regions that represent words and emotions may be wider than the halo of mathematicians. The halo of a car-mechanic is wider where information about cars is stored.

The extent of the halo is dynamic; it may change according to the actual information that is being processed. It may depend also on conscious and subconscious associations that are triggered in the thinking process. The halo contains nodes that are already active or are the first to be retrieved in the ongoing mental processes.

THE ROLE OF ITEMIZATION AND CLASSIFICATION DETECTORS

Morie

Since itemization and categorization rely on the available detectors in the brain, and since most of the detectors are learned, the overall intelligence of a person would depend on the detectors that have been developed in that person's brain. Many new detectors require repeated experiences in order to assemble their cues. Intelligent brains assemble such cues after less repetitions. In the same amount of time, they develop more detectors than the less intelligent brains. They have more cues with which to itemize and categorize items from the information flow, thus they understand it better. As is commonly said, less intelligent individuals are "clueless".

Those differences in performance, which are at the level of nodal mechanisms, mainly affect understanding at the *knowledge* and *comprehension* levels. At higher levels of understanding, the functionality of the thinking apparatus is an important factor in the individual's intelligence. Operations at those levels require the coordination of many basic mechanisms and the evaluation of partial outcomes. Those activities are handled by the thinking apparatus.

THE THINKING APPARATUS

Morie

When given a difficult thinking task, the thinking apparatus divides it first into small sub-tasks, and expresses each sub-task as a logical combination of basic mechanisms. It then activates the first sub-task, while keeping the others at the background. Once the first sub-task is completed, its outcome is evaluated by the thinking apparatus, which then activates the next sub-task, and so on, until the task is done.

The thinking apparatus of intelligent people can coordinate the activities of more sub-tasks in more complex configurations. It can keep sub-tasks in a dormant state until they are needed. It then can activate them according to the outcomes of other active sub-tasks. It can store partial outcomes and merge them at the appropriate time with the appropriate sub-tasks. Those capabilities of the thinking apparatus, which affect the overall intelligence, vary significantly between individuals.

Claire

Can we see an example that illustrates how the thinking apparatus coordinates sub-tasks?

Morie

We do not have to go far. Let us look at how typical people calculate in their head a simple multiplication problem: 53 times 7. Many people would do it like this: First, express 53 as 50 plus 3. Then do 3 times 7 equal 21. Next do 50 times 7 equal 350, and finally do 21 plus 350 equal 371.

The thinking apparatus first classified the task as an exemplar of the class 'multiply a two-digit number by a single-digit number'. This task is already expressed as an assembly of sub-tasks in our brain. The first sub-task is to divide the two-digit number into tens and ones, and remember each. Then multiply the ones by the singles digit and remember the outcome as the first part of the final answer. Next multiply the tens by the singles digit, add the result to the first part of the final solution, and call it 'the answer'.

This entire chain of operations consists of retrievals ('50' and '3' are retrieved from '53', based on their feature of being the first and second digits of a two-digit number.) The item 21 is retrieved based on its parts '3', '7' and their relationship 'times', similar to the example of Figure 8. The item '350' is retrieved based on '50', 'times', and '7'. Adding 350 plus 21 is done by another sub-task, which has its own sub-tasks. The thinking apparatus organizes the operation of all the sub-tasks, and shuttles information back and forth between the sub-tasks and the memory, as needed.

Claire

So, when we do complex calculations on paper, the paper acts as an extension of our memory, and the thinking apparatus treats it as an extension of the brain.

Morie

This is an interesting perspective. Paper and, more recently, various digital storage media could be viewed as extensions of human memory. Calculations done by computers may be viewed as extensions of the human thinking apparatus.

Sol

How does the thinking apparatus divide a task into sub-tasks?

Morie

It has various activity-plans for doing that. In general, it would try to identify parts that make up the entire problem. In the previous example, it would divide the number into tens and units, and work with them instead of the entire number. Another powerful approach is to look for classes that contain as exemplars the problem, or parts of it. In the previous example, the key step was to realize that the problem is a special case, in other words an exemplar, of the more general concept, in other words a class, of how to multiply two numbers, and then to recognize the role of each digit. That means to treat each digit as an exemplar of a class. By doing that, each digit inherits the common properties of exemplars of the same class. In this example it was how they interact with other digits. Intelligent people see the relationship between the problem at hand and general situations, that may seem at first unrelated. In our formulation, intelligent people can assign exemplars to classes based on few and weak cues.

THE SAVANT SYNDROME

Bob

There are savant people that can do complex arithmetic calculations in their heads. Can our model explain that phenomenon?

Morie

The Savant Syndrome is a rare and spectacular condition in which people with various developmental disorders, such as autism, show amazing capabilities in limited specific skills. Not all autistic individuals are savant, and not every savant is autistic. In the movie Rain Man, Dustin Hoffman plays the role of Raymond, an autistic savant, who can calculate complicated mathematical problems in his head with great speed and accuracy. This is the savant component of his condition. His autism is expressed by the ways that he conducts himself in his daily life and by how he perceives and grasps his environment. For example, in spite of his amazing computational skills, he does not understand how money is used.

Savants have shown astonishing skills in music, arts, languages, accurate intuitive determination of passing time, lightning calculations, and more. However, their performance outside the boundaries of their specific skill is usually very limited. For example, a savant that could calculate in his head multi-digit prime numbers had difficulties in solving simple arithmetic problems.

It is not known what causes the Savant Syndrome and what are its underlying biological mechanisms. However, it could be explained by the basic nodal mechanisms described here. Due to one reason or another, when dealing with the Savant Syndrome and autism, those mechanisms apply in only limited parts of the brain, and are absent in others. The brain's detectors of a certain savant may detect numbers and words and record them in the memory very effectively, so that long phone directory lists could be recorded and retrieved with amazing efficiency by that savant. At the same time, some detectors, whose development depends on innate emotion centers, e.g. those involved in identification and communication with other humans, may not have evolved properly due to impairments in those centers. Impairment of such emotional centers affects the development of new item and class nodes that are crucial for the interpersonal and social interactions of the individual.

MATH WIZARDS, HISTORY BUFFS, AND SPORTS MAVENS

Bob

I was never very strong in math, to put it mildly, but I excelled in history. When it comes to sports trivia, many call me a walking encyclopedia. How all that relates to the nodes and the synapses in my brain?

Morie

It all starts with how effectively the brain records and organizes the information that it will use later. The emotions associated with an information item affect the rate of its recording. Emotional information is recorded faster than emotionless information. My guess is that you are more excited about sports events than about math theorems. That increases the rate at which synaptic weights are modified when sports information is recorded. An unexciting math theorem will have to be repeated more times, until the small changes in synaptic weights that each repetition induces will be sufficient to establish the brain record that represents that theorem.

Various emotions may affect the recording rate of new information. For example, distress due to an unpleasant experience increases the rate that it is recorded. This is important for survival, because it enables the organism to avoid repeated experiences of harmful or unpleasant situations. The other side of the same coin is that also pleasant emotions increase the recording-rate of their associated events. Since you were exposed to both history and math at school, you may have more pleasant associations with history than with math. By its nature, history is more about human emotions than math. That may have given history a slight initial advantage. Then, you probably were rewarded more by your initial success in history than by your struggles with math. Your teachers praised you, and your classmates appreciated your good grades. That added to the emotional superiority of history over math. When watching sports, you probably have identified yourself with a team, and may have developed emotional attitudes towards other teams. You were looking forward, full of hopes and concerns, to games between your favorite team and its rivals. You were also very emotional after the games, whichever way they went. All those emotions, which expanded also to other sports topics, made it easier for your brain to record that kind of information.

The education system is aware of the importance of emotions in learning. Grades are means that the system uses in order to associate emotions with emotionless subject matters, which, if given the choice, students would rather not take. The desire to succeed and to be appreciated by teachers, parents, and peers enhanced the learning process of topics that became your favorites.

In addition to the differences in their emotional overtones, other intrinsic features make math more difficult to master than history, admittedly for many students. As we have seen, learning involves two basic processes: itemization and classifications. In effective learning, items, which are the "main players", are defined, and then they are classified, so that their intrinsic and inherited properties become available to the brain's mental operations. The items in history are easier to assemble than the items in math. Many students, when they hear it for the first time, would find that defining the item-node "Julius Caesar", based on its features "ruler", and "Rome", is easier than defining the item "the square root of X" based on its features: a number that when multiplied by itself gives X.

Once a concept is established as a node connected to other nodes, the level of its usefulness depends on the number of its inherited properties that are readily available to mental operations. In most brains, 'Julius Caesar' would have more inherited features readily available for mental operations than 'the square root of a number X'. All the inherited properties of a human being and of political life are readily available with 'Julius Caesar', and could be used in understanding his life. 'Julius Caesar' is an exemplar of 'ruler'. As such, it is obvious that he was involved in politics, sought and enjoyed power, and had supporters and enemies. All these concepts are easily accessible, and make it simple to understand why Brutus assassinated him. On the other hand, when asked to find the square root of four, a novice may answer: "two, because two times two equals four". When pressed "is this the only number?" the novice, who hears the question for the first time, might answer "yes". When asked "what about minus two?" the response would be: "oh, right". The fact that minus times minus is plus, although known to the novice, is not readily available in his brain. Therefore, the novice could not get the answer that minus two is also a square root of four.

History is more related to the human experience than math. Once the items of history are defined, they enjoy the existing rich connectivity with the rest of the brain's information structures; a situation that many math concepts do not enjoy.

Bob

In my brain, the halos of sports trivia and history items are wider than the halos of math problems.

Morie

That is a bright observation. And, of course, the thinking apparatus of some people is more suited than that of others to handle math problems. They are better at uncovering and using inherited information, coordinating operations of sub-tasks, and shuttling information between them.

Claire

To me, math did not come easy either. I used to sit long hours before each test and solve problems, memorizing many of them. I got good grades.

Morie

By solving many problems, your brain has benefited in two ways. First, it developed more detectors that apply to math problems. Your brain identified common features that repeated in the problems, and made them cues for new classes. When a new problem was presented, those detectors could identify relevant features that existed in the problems, and use them in the solution. Those detectors enhanced the halo of the original nodes that represent the problem.

The second benefit was that problems at the levels of *analysis* and *synthesis* of Bloom's Taxonomy became problems at the lower levels of *application, comprehension* and *knowledge.* Any problem that required analysis and synthesis, which you solved at home, and then luckily showed up in the test, was just a matter of knowledge; you had to retrieve the solution. If you solved at home a similar problem, the test problem did not require *analysis* anymore; it was just a matter of *knowledge* or *application*, which utilize the simpler mental operations of switching exemplars between class nodes.

IN A NUTSHELL

Claire

Allow me to summarize very briefly the main point of our discussions so far. The innate brain is a network of nodes connected by synapses. The nodes represent concepts, and the connections between them encode basic relationships between concepts.

Some of the synapses are fixed and others are plastic and may be modified. Information is recorded in the brain by the modification of synaptic weights. This is the underlying mechanism of learning. As individuals interact with their surroundings, the brain records those experiences in interconnected hierarchies of item-parts and class-exemplars.

Nodes act as detectors; they fire in response to certain patterns of incoming information. The firing of some detectors initiates synaptic modification in other nodes that have been involved in that firing. Detectors are closely involved in itemization and classification of information and in the construction of the item-parts and class-exemplars information hierarchies.

Detectors are also involved in creating other detectors. New detectors that are added to the brain are produced by existing detectors as they process incoming information.

At the nodal level, human behavior is determined by the innate detectors of the brain, by acquired detectors, by the inter-connected items-parts and class-exemplars nodal hierarchies that represent the brain's stored information, and by the interactions of all these nodes with external stimuli. The basic operations at the nodal level, which support all higher-level mental activities, consist of manipulations of connections between nodes, and information retrieval requests, expressed as combinations those basic requests.

Any problem that the brain is solving is represented by active nodes. In addition to the original nodes that represent the problem, there is a halo of nodes, which are associated with the original nodes. The halo consists of firing nodes and primed nodes, which are partially activated. The extent of the halo is one of the factors that affect the quality of the solutions that the brain could find to the problem.

The brain has a thinking apparatus that organizes and coordinates the execution of complex tasks. The thinking apparatus divides a complex task into sub-tasks. It manages the proper activation of the sub-tasks and the exchange of information between them and the memory.

DRIVES

DRIVES AND ACTIVITY-PLANS

Morie

Triggered detectors provide the brain with information about the external world and about our feelings and emotions. Similarly, our actions are triggered by activators. A basic activator is an innate node that activates a muscle unit, a group of muscles, or a gland. Acquired activators are combinations of basic activators. Figuratively, activators may be viewed as outlets of information streams, whereas detectors are inlets.

Similarly to detectors, two main mechanisms underlie the formation of acquired activators. First, repetition. A pattern of activators that fire repeatedly coalesces and forms one new activator. The new activator is more self-contained than its constituting activators. It needs less external controlling signals.

For example, many movement activities such as dancing, riding a bike or driving a car are assembled from basic activities. At first, the coordination between the basic activities requires our full attention. After repetition, the basic activities join to form one entity that requires much less attention.

In the second mechanism, patterns of basic activators that invoke strong rewards or negative repercussions are acquired after only one or a few repetitions. For example, it does not take long for a baby to learn to scratch an itchy spot. Apparently, the gratification generated by the scratching accelerates the acquisition of that habit.

The nodal mechanisms that underlie the acquisition of new activators are the same as those for detectors: recruiting new nodes and modifying synaptic weights.

Most of our non-trivial activities are executed by activity-plans. Activity-plans combine detector and activator nodes. When triggered, the node that represents the activity-plan activates or primes its subordinate detector and activator nodes. As the activity progresses, detectors trigger activators according to the progress of the activity.

For example, when we decide to brush our teeth, the node that represents that activity fires. It primes the detectors of the toothbrush and the toothpaste, and activates the muscles that move our head and eyes in searching. Once the brush and the paste have been detected, the activators that move the hand are triggered, and so on.

Claire

Some of our activities appear to originate from inside us; they do not require external stimulation. What drives them?

Morie

Several types of activities are driven from within. Activities of autonomic vital functions, such as breathing and digestion, are controlled by innate activity-plans. Those activities are adjusted by detectors in response to external conditions.

For example, breathing is driven by certain nodes, but its rate and intensity depend, among other things, on blood oxygen levels. Breathing rate and intensity are adjusted by designated sensors and detectors, as part of the overall activity-plan. Such activity-plans are innate, and they cannot be modified. Some of them, such as regulating blood circulation, are involuntary, while others, such as breathing, have also partial voluntary controls.

HOMEOSTASIS

Morie

Homeostasis is the term used to describe the tendency of many biological systems, including the human body, to maintain a state of equilibrium. Homeostasis requires the coordination of body activities with prevailing external conditions. The role of several basic drives is to maintain the equilibrium of the body.

For example, when detectors detect that it is too hot, the blood-sugar-level is low, a body part is squeezed, and so on, a corresponding activity-plan is triggered. Generally, the triggered activity, e.g. moving or crying, would continue until the detectors do not detect the offensive condition any more.

The concept of homeostasis could be extended also to emotional equilibrium. Innate drives tend to maintain us in emotional equilibrium. Like physical sensations, which are pleasant or unpleasant, there are pleasant and unpleasant emotions. The brain has innate drives whose aim is to maintain pleasant emotions.

For example, the sensation of soft, warm body contact creates in a baby the feeling-safe emotion, and the baby is relaxed. If that emotion is disrupted, another activity, such as crying, is triggered, until the emotional homeostasis is regained.

Overall, homeostatic-needs drive activity-plans. The activity-plans are elements of feedback loops that maintain the equilibrium of the system.

In addition to drives of vital functions, the brain has other innate drives. Locomotion is one of those drives. Many animals have a drive to move around just for the sake of moving; they roam, fly, or swim even when they are not hungry or threatened. They feel uneasy if their freedom to move is restricted. Once such a restriction is lifted, they resume moving and feel relieved. Moving around would benefit them indirectly, such as by finding new food sources, but the lack of such benefits is not what drives them to move in the first place.

Another basic drive is the need to get stimulated. Humans of all ages seek novel stimulations. If the surroundings do not provide enough of it, the bored brain would trigger activity-plans such as moving, handling objects, talking, watching TV, playing games, and the like, whose purpose is to generate stimulation.

In addition to the vital and homeostatic drives, there is another group of drives, which are often called hedonistic. Whereas homeostatic drives tend to maintain the body at a comfortable equilibrium state, hedonistic drives aim to increase gratification and pleasure sensations beyond the equilibrium level. Hedonistic drives are learned. The brain keeps records of activities that have generated pleasure. Such combinations of an activity and its ensuing pleasure may have occurred randomly for the first time, but after they have been recorded, the brain uses the records to drive again those activities, expecting a repeat of the associated pleasures.

For example, people who have never tasted chocolate will not look for it, but some of those who have tasted, crave it. They have an acquired drive to get it.

An activity-plan may be activated by more than one drive. For example, looking for a companion may be initiated by the emotional drive 'to feel safe' and by the drive 'to seek stimulation'.

Claire

I must tell you about Mitzy, my cat. It is cozy at home, plenty of cat-food and cat-toys. But Mitzy has to go outside, sniff objects, crawl into the street's drain system, come out from the other side, return home, yawn and take a cat nap. She has drives to move around and to get new stimulations.

Bob

She may also have a drive to find a Mickey.

EXPANDED DRIVES

Morie

Drives may be divided into two groups from a different perspective: drives with open-ended activity-plans and drives with goal-oriented activity-plans. An open-ended drive triggers activities that do not have explicit goals, other than the activity itself, e.g. locomotion. A goal-oriented drive has a well-defined goal, such as getting food.

Both kinds of drives could be expanded by learning. New triggering cues could be added to the activity-plans, and new activity-plans could be added to the repertoires of the drives. Goal-oriented drives could be expanded also by adding new goals to their activity-plans.

For example, throughout their entire life, humans acquire new cues that would trigger their sexual arousal. Those cues are acquired through physical experiences and by mental processes.

Throughout history, individuals have been developing a wide variety of courtship strategies. Romantic poetry and "pick up lines" are just two examples from the expanded repertoire of courtship activity-plans.

In order to push merchandise, advertisers use methods that expand the goals of sexual drives. One genuine sexual goal could be stated bluntly as 'get hold of a sex object'. Advertisers expand it to 'get hold of something that is associated with a sex object'. By blending sexual motifs, advertisers turn asexual objects, such as cars, into targets of sexual activity-plans. The advertised car with a beautiful supermodel at the wheel becomes a target of a sexual activity-plan. The potential buyer wants to get hold of it. At the end, the buyer does not even think of suing the advertisers for switching bait.

OPERANT CONDITIONING

Sol

What could be the underlying mechanisms by which drives are expanded?

Morie

Drives are expanded like any other concept – by the recruitment of new nodes and the modification of synaptic weights. Operant conditioning is a common process for expanding the activity-plans of drives.

It is possible to train a mouse to push a lever that dispenses food. Initially, the mouse will push the lever randomly. The dispensed food rewards the activity, and eventually the mouse would push the lever whenever it wants food. Pushing the lever is an activity-plan that was added to the repertoire of the mouse's 'get food' drive by operant conditioning.

Claire

What is the difference between operant conditioning and classical conditioning?

Morie

In classical (Pavlov's) conditioning, the subject's response to the stimulus is involuntary, e.g. salivation, whereas in operant conditioning the response is voluntary.

Figure 13 illustrates an underlying nodal mechanism of operant conditioning

There are several variations of operant conditioning. In one version, the subject is rewarded when it chooses the targeted response, and penalized when it responds otherwise. With repetitions, the activity-plan of the targeted response is reinforced and the others are weakened.

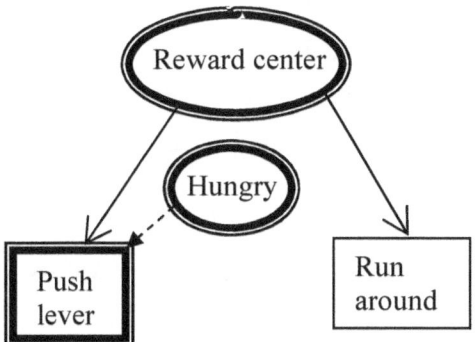

Figure 13 A nodal mechanism of a mouse learning to push a lever by operant conditioning. Highlights indicate firing nodes. The reward center is triggered by the sensation of the food that has been dispensed by the pushed lever. The reward center broadcasts to many activity-plans (rectangles). A connection is formed from the 'hunger' detector to the 'push-lever' activity plan.

It is also possible to reward the targeted response and to ignore, instead of penalize, all the others. This approach is called positive-reinforcement operant conditioning.

It is possible to create an activity plan by associating new cues with an innate activity. For example, when training dogs, saying 'sit' and gently pushing the dog down, to initiate its normal sitting activity-plan, will train the dog to sit when it hears those words. The dog would sit not only when it is tired, but also when it hears those sounds. At the beginning, the dog is rewarded by food after it sits. With repetition, the dog would sit when it hears the instruction, even without the rewarding food.

Operant conditioning may be applied to expand the goals of a drive. If an arbitrary cue is presented in conjunction with the reward, this cue might become a goal of a new drive.

For example, when the reward is food, dog trainers use a clicker or a whistle, and some just say "gooood dog" after the dog has behaved as expected, and before rewarding it with food. With time, such cues become independent rewarding goals for the dog. Saying "gooood dog" could be used as a reward in training the dog by operant conditioning. The dog's desire to hear "gooood dog" may be considered a new drive, or at least an expansion of existing ones.

ORGANIZATION OF BEHAVIOR

ACTIVITIES-COORDINATOR

Sol

How does the brain organize the activities of all its available drives and the other activity-plans?

Morie

Let us restrict our discussion to voluntary activities. The brain has a system whose role is to select and coordinate the activity-plans that are running at any given time. We will call that system the 'activities-coordinator'. The activities-coordinator receives requests for activities that originate from drive-nodes and from other sources, such as verbal communication. One common form of such requests is 'an action is needed to achieve this given goal'. For example, 'an action is needed to get to work', or 'an action is needed to have some fun', and so on. The activities-coordinator also receives stimuli about the current state of the body and the environment. It first retrieves an activity-plan that matches the prevailing conditions and the given goal. It then may run a mental simulation of the retrieved activity plan. If the detectors do not detect any possible adverse outcomes that call for vetoing the activity-plan, it is activated. Otherwise, another activity-plan, which takes into consideration the shortcomings of the original plan, is retrieved.

Finding an activity-plan relies on the same thinking mechanisms that retrieve information based on given cues. There are several sophistication levels in finding an activity plan, as we detailed in our discussion of Bloom's Taxonomy. For example, in synthesis, new activity-plans are constructed from existing ones, to answer the specific goals. The highest level is evaluation, in which competing activity-plans are compared and evaluated.

One of the roles of the activities-coordinator is to prioritize competing activity-plans. In some cases, a priority scale that depends on the given cues, the activity-plans and the target already exists, and the activities-coordinator relies on it (e.g. should I start with the meat or with the potatoes? Most people would start with the potatoes.) When an explicit priority scale does not exist, selecting an activity-plan could be accomplished by a thinking mechanism, e.g. should I have ice cream or cake for desert? Oh, the D word! I am on diet. I'd better have a coffee. That decision involved high-level processes on Bloom's Taxonomy – analysis and evaluation.

Claire

Is it known where in the brain is the locus of the activities-coordinator?

Becky

The prefrontal cortex is implicated in most of those activities. It is yet to be uncovered how the different tasks of the activities-coordinator are performed on the cellular and circuit level. However, lesions to the prefrontal cortex disrupt those activities. In addition, the time-line in which children acquire the ability to evaluate activity-plans matches the time-line of the maturation of the prefrontal cortex. That includes the sequence of arborization of its dendritic trees and the myelination of its axons. In addition, centers in the pre-frontal cortex show activities that correlate with anticipation, an important element in planning and coordinating sequences of activity-plans.

EXAMPLE

Sol

Could you give an example of how the activities-coordinator operates?

Morie

When we are hungry, the activities-coordinator has to retrieve or construct an activity-plan that will end our hunger, taking into consideration our location. Assume that we are at home and we want to eat as soon as possible.

The activities-coordinator activates a master activity-plan to deal with hunger, which has several sub activity-plans. The first to be activated is 'where is food available?' This is a retrieval request for information stored in our memory. It would retrieve a number of nodes such as 'in the refrigerator', 'in the store' and 'in the restaurant'.

In order to evaluate them, the activities coordinator would initiate three mental simulations. Each retrieved option of food location would be provided as input to the activity-plan of 'go from where you are to where the food is'. Detectors in the brain would be activated in those simulations, and one of the detectors would detect how long it should take to get the food. The shorter time needed to get to the refrigerator would trigger a feeling of satisfaction, and would single out the refrigerator as our chosen destination. The activities coordinator would pipe this choice to the next activity-plan 'go to the refrigerator', activate it, and so on.

ACQUIRED DRIVES AND ADDICTION

Bob

What are the underlying brain mechanisms that make craving for chocolate so different from becoming addicted to drugs?

Morie

Both start with the same underlying mechanisms. During the first experiences, the onset of a pleasant feeling enhances the recording of the event. After that, when a need to repeat the pleasant feeling arises, the brain retrieves the associated activity, and if approved by its activities-coordinator, the person eventually would get the chocolate or the illicit drug, as the case may be.

However, in addition to being recorded like any other event, many illicit drugs interact in a variety of ways with other components of the brain. For example, persistent changes in blood flow and in glucose utilization have been documented in brain regions of amphetamine and cocaine addicts. Decrease in the volume of the pre-frontal cortex has been documented in addicts that abuse more than one drug. Chronic amphetamine and opiate users show deficits in decision-making abilities and in their ability to inhibit or to properly assess the consequences of their action. The latter impairments are typical to patients with lesions in the pre-frontal cortex. Such side effects influence the general function of the brain, including activities unrelated to drug use.

Unlike chocolate, drug users develop tolerance to drugs; they have to increase the amount of drug needed to satisfy them. Another side effect that distinguishes drugs from chocolate is withdrawal symptoms. Drug users may develop dependency on the drug, so that if they try to stop taking it, agonizing and painful side effects follow.

Micropsychology

The presence of drugs in the brain interferes also with synaptic mechanisms and in general communication between neurons. The result of one of those interferences is sensitization. Repeated use of drugs can intensify arousal, attention, locomotion, exploration and related activities. Those changes may be attributed in part to strengthening of connections between nerve cells that participate in those activities.

Bob

Have such changes been really observed in neurons?

Becky

Yes. Cells in the nucleus accumbens and the prefrontal cortex show morphological changes that result from using amphetamine, cocaine, or morphine. Changes were observed in the length of the dendrites, in the amount of their branching and in the numbers of their dendritic spines. Those cells are parts of neural circuits that release neurotransmitters associated with reward. Those changes may affect the reward mechanisms, which in turn affect the recording of events associated with the drug use.

Bob

At a certain point, drug addicts keep taking drugs even though it hurts them, and they are aware of it. Some drug users say that they do not enjoy it, but they must take it. Why is that?

Becky

In a series of controlled experiments on mice, sugar was used as a reward in operant conditioning. Mice learned that if they press a lever when a light flashes, they are rewarded by a drop of sweet water. Three groups of mice were tested. The first was the control group. The second were administered amphetamine before the training. In the third group, amphetamine was microinjected directly into the nucleus accumbens during the training. It was found that the drugged mice developed greater propensity to pressing the lever, compared with the control group. After the training ended, they pressed the lever even when they did not want the sweet water. The presence of amphetamine in their brains elevated the priority that their brains assigned to the ongoing activity – pressing the lever.

This suggests that in addition to the hedonistic incentive of drugs, which are acting like any benign hedonistic agent, drugs affect the priority that the brain assigns to taking drugs. This activity gets a higher priority than other activities, and it becomes a goal in and of itself, even if it does not create pleasure anymore. When the activities-coordinator has to choose the next activity, taking drugs has a high priority. It would take very little, in the form of any cue remotely associated with drugs, to bring that activity to the top of the list. That may be why it is so difficult to stop that behavior, and why recovering addicts relapse.

CONCLUDING REMARKS

THE PRINCIPLES

Claire

The picture that emerges from our discussions is that a relatively small number of basic mechanisms and principles at the cellular level serve as the building blocks with which we can describe many aspects of human behavior.

The neuron can be modeled as a basic information-processing element. It has input channels, integrating body, and output channels. Signals enter through the input channels and are integrated. If a certain threshold is passed, the neuron "fires"; it sends signals through its output channels. Usually, a neuron would communicate with many other neurons through its input and output channels.

Neurons are divided into three groups, according to the information that they relay: Input neurons relay information from outside the brain into the brain; output neurons relay information from the brain out, and inter-neurons, the vast majority, communicate among themselves and with input and output neurons.

Information is represented in the brain by nodes that are connected in networks. All the operations that the brain performs as it processes information can be described by nodes and interaction between them. A node is a group of neurons that represents a concept. Like neurons, nodes have input channels, output channels, and a computation-unit. The computation-unit of a node integrates the node's input signals. The node fires if a threshold is passed. Unlike neurons, communication between nodes could be bi-directional.

New information is recorded in the brain by processes that modify the synaptic weights between neurons. In some processes, such as imprinting or those that involve drug abuse, the synaptic changes cannot be reversed. In other processes, the synapses do not lose their plasticity, and the processes are reversible. In some cases, the brain may annul acquired information by developing information pathways that bypass it.

Four basic relationships between concepts are represented by the weights of the synaptic ties of the nodes that represent the concepts. The relationships are: A is an exemplar of B, which is the same as B is a class of A. These two relationships are called 'is-a' relationship, as in 'Tom is a boy'. The other two relationships are: A is a part of B, which is the same as B is an item of A. These two relationships are called 'has-a' relationship, as in 'Tom has a nose'.

Consequently, information is represented in the brain as interconnected 'is-a' and 'has-a' hierarchies.

Functionally, a node has two parts: a detector and an activator. The detector causes the firing of the node when a certain input pattern enters its input channels. The activator sends signals to other neurons and may initiate certain activities, such as recording the event. In many cases, a detector would respond to more than one specific pattern, and the activator would need cooperation of other activators in order to trigger its target nodes.

The innate detectors and activators determine how we perceive the world. They build up our perception from innate senses and feelings, using concepts such as 'the same', 'change', 'before', and more.

Activators trigger a variety of activities. They activate the muscles and the glands and modify the hardware of the nodal network. Those modifications are important stages in recording new information or retrieving stored information.

The brain expands its functionality by learning new information through the modifications of the weights of synapses between its nodes. The brain learns new detectors, new activators and new data, and uses them together with the innate ones.

Genes determine the structure of the brain, the innate information that it would have, and the basic information processing mechanisms that it would use throughout life. The experienced brain is the outcome of its genetic make-up and the external events that it has been encountering. What we are and what we do could be traced back, in principle, to our genes and our experiences.

Retrieval-requests are processes that underlie many mental activities. Retrieval-requests may be simple or compound, but they could be expressed as logical combinations of four basic retrieval requests: 'retrieve a class of an active node'; 'retrieve an exemplar of an active node'; 'retrieve a part of an active node' and 'retrieve an item of an active node'.

When executing a compound retrieval request, the brain can evaluate its subparts and adjust the process according to partial outcomes. Such adjustments are expressed by replacing problematic portions of the request. The replaced portions are also expressed as logical combinations of the four fundamental requests. Retrieval requests may result in the creation of new information that has not been experienced before and that is not recorded in the brain.

The brain has an activities-coordination unit, which coordinates its activities. The unit sets priorities and regulates the activities that span the behavior of the individual. The regulation of activities is also based on retrieval mechanisms that employ the basic retrieval processes.

SOME IMPLICATIONS AND APPLICATIONS

Bob

This has been a nice summary of the principles of our theory. Would it be possible to show some implications and applications of it?

TEACHING

Morie

I can give some; the first is from the area of teaching. In teaching, a teacher relays information to a student. That information has to be incorporated in the student's brain so that it could be used later, together with previously stored information. The relayed information is represented in both the teacher's and the student's brains by nodes and synapses. The goal is that the relayed information-structure in the student's brain emulates its source in the teacher's brain.

For this to happen, relayed item nodes and class nodes in the student's brain should have the same cues as their counterparts in the teacher's brain. They should also have matching connections with the rest of the nodes of the information structures, which were not relayed by the teacher.

For example, consider a teacher teaching a student about triangles. Assume that initially some concepts such as 'angle', 'degrees', 'right-angle', 'acute angle', 'hundred and eighty degrees' already exist in both of their brains. Assume that the student is not familiar with triangles. The goal of the teacher is to induce in his student's brain nodes and connections that represent information about triangles.

The teacher draws a triangle while saying what he is doing. Detectors in the student's brain process the streams of incoming stimuli, and record all this sensory information as nodes and connections. An item node that represents the new concept 'triangle' is recruited. Nodes that represent the sides and angles of the triangle and their inter-relationships are recruited and become parts of the item node 'triangle'. The word 'triangle' is also represented by a node in the language areas of the brain, and the sensory node 'triangle' becomes one of its exemplars.

Then, the teacher draws a right-angle triangle, saying what he is doing. This stream of stimuli is analyzed by the detectors of the student's brain and recorded. The newly formed detectors of the concept 'triangle' and its parts, together with older detectors, such as 'right-angle', participate in the process. The teacher then asks the student to draw a right-angle triangle, and she draws it. That is an indication that the new information has been incorporated in her brain's information structures. Her brain can access the information and use it in its operations.

Now the teacher asks "what kind of angles do you see in the triangle that you drew?" The student answers "one right angle and two acute angles". Her brain emulates the teacher's brain and retrieves the correct answer.

The teacher goes on: "if one of the acute angles is thirty degrees. How many degrees is the other one?" The student thinks and says "I don't know". There are differences between her new information structure and that of the teacher. The teacher has not relayed some of his information-structure that is needed for answering that question. In his brain, the class node 'triangle' has a part node that represents a property shared by all triangles—the sum of their degrees is hundred and eighty. This crucial piece of information was left out in the teaching, and the student could not generate it from her information structures.

Usually, relayed information has implicit and explicit parts. The implicit parts are not relayed directly—the students are expected to be able to deduce them from information that already exists in their system. However, they could do it only if those parts really exist in their system, and if those pre-existing parts have been connected properly with the new information.

The stream of information that originates from the teacher is processed by the student's detectors that guide learning-activators in the incorporation of that information. In a successful teaching, the relayed cues, parts, items, exemplars and classes are replicated in the student's brain. The connections that their nodes form with other nodes are similar to those in the teacher's brain. In a partially successful teaching, some of those elements have not been replicated properly—representations of some information entities are missing or misconstrued. Such deficiencies show up as the student not understanding the material. To correct it, the teacher should identify the source of the problem. Is it because some cues are missing? Is it because a concept is not represented at all? Is it because the right connections to pre-existing nodes were not established? Is it because some deficiencies of the pre-existing student's information system? Pinpointing the problem is the first step of correcting it.

These kinds of problems affect all levels of understanding, as classified by Bloom's taxonomy. Problems in the levels of analysis, synthesis and evaluation may arise because of pre-existing deficiencies in the student's system, even if all the information has been relayed and recorded properly in the learning process. Those deficiencies involve activity-plans of thinking processes, which are required for handling the recorded information. In order to correct for such deficiencies, the teacher may have to go beyond the boundaries of the new subject matter that has to be taught.

The detectors of the brain execute the learning in a mechanistic way. They identify in the information stream information entities and their types, and configure them as representations of nodes and connections. For example, the teacher does not have to say "the triangle that I am drawing is an exemplar of a class, whose cues are three lines, three angles, etc". The brain's detectors do it automatically. However, the teacher could facilitate the task of the detectors by presenting the information in a way that matches their mode of operation.

For example, when introducing triangles for the first time, the teacher could draw a few different looking triangles. The detectors would find the common features to those shapes and use them as cues of the new class 'triangle". This is how detectors operate naturally. They learn from examples.

The brain's learning mechanisms define subclasses by comparing and contrasting information entities. Distinction between subclasses is important to understanding the material. The teacher should make sure that appropriate examples are provided, so that the brain can identify the subclasses and their cues.

The teacher could also describe in words how lines and angles make up the triangle. That would highlight the features of the new item, and facilitate their incorporation as cues by the brain's detectors. In general, explaining the relationships between information entities, e.g. pointing to part-item relationships between new concepts, facilitates their incorporation in the network.

ADVERTISING AND PUBLIC RELATIONS

Morie

Like teachers, advertisers and PR professionals relay information that has to be incorporated in the receiver's brain. Quite often, the exposure time of the receiver and the message itself are shorter in advertising than in teacher-student settings. Because of that, advertisers have to tailor their messages more efficiently, and invoke in the receiver efficient recording mechanisms.

A general strategy of advertisers is to associate the subject of their campaign with a specific node that has many, inheritable properties. The only change that the advertisers have to induce in the receiver's brain is a connection between two nodes. Once that connection is established, everything else follows automatically.

For example, it is now common to discredit candidates in an election by drawing attention to negative traits that they have. Those traits serve also as cues to classes with derogatory inherited properties. If a candidate is a slow speaker and an opponent succeeds in making it a defining cue of that candidate, he would be classified in the voters' brains as an exemplar of the class 'slow speakers'. He would inherit properties of that class such as boring, ridiculous, slow thinker, dumb, and so on, which could derail his candidacy. All the opponent's camp has to do is to emphasize repeatedly that the candidate is a slow speaker. Once the candidate is perceived as such, the rest would be done by the voter's basic brain mechanisms.

The same strategy is used also with positive traits of humans and inanimate products. 'Champaign' has become one of the inherited properties of the class 'celebration'. Any cause for celebration that happens to us is encoded in our brain as an exemplar of that class, and invokes the inherited property 'Champaign time'.

A basic mechanism of the brain is to associate events that coincide repeatedly. If those repetitions are accompanied by an emotion, they would be recorded even faster. Advertisers keep repeating their messages in emotion-stirring situations. The brain learns faster events that are associated with fear, joy, envy, and so on.

The brain does not filter the stream of incoming stimuli in order to avoid learning any false information. Rather, identifying recorded information that is false is usually a separate process, and if it is done, it is done after the recording process has ended. Therefore, it makes sense to ask ourselves from time to time, "How did I get this idea? Is it a valid idea?" False perceptions may be lurking in our minds. Identifying them would help us to get rid of them and reach valid decisions, because quite often, the brain does not check for those flaws unless prodded.

SEXUALITY

Morie

In order to manage the physiological machinery of sexual arousal and attraction, the brain relies on information that it has learned. Starting in childhood and throughout life, that information is learned automatically without us being aware of the learning process. The same brain mechanisms that acquire, encode, record, and retrieve sexual information are used by the brain to handle all other type of information.

Arousal and attraction information are recorded in hierarchical structures, at the top of which are class nodes that activate the physical arousal and attraction mechanisms. Those class nodes are triggered by cascades of exemplar nodes that represent different arousing situations. The exemplar nodes have evolved as a result of individual experiences, and they embody the sexual preferences of the individual. Each exemplar node has detectors that detect specific feature-combinations in the information flow. Those feature combinations indicate the presence of an arousing factor. Once such a factor is detected, activation propagates up in the information structure, until it reaches the top and activates the physical arousal mechanisms.

Parts of the information-structure are purely genetic, and parts depend on individual experiences. Figure 14 illustrates the information structure of sexual arousal and the direction of information flow in it. It also highlights the learned parts of the information structure. A similar information structure handles sexual attraction.

Two main learning mechanisms build up the information structures. The first assembles parts into items and the second assigns items as exemplars to classes. The first mechanism is responsible for defining what a concept is made of, in other words the 'has-a' of a concept, e.g. the tree has leaves. The second mechanism is responsible for defining to what group of similar concepts the particular concept belongs, or in other words, the 'is-a' of a concept, e.g. a pine tree is an evergreen. Both mechanisms take part in building the sexual information structure.

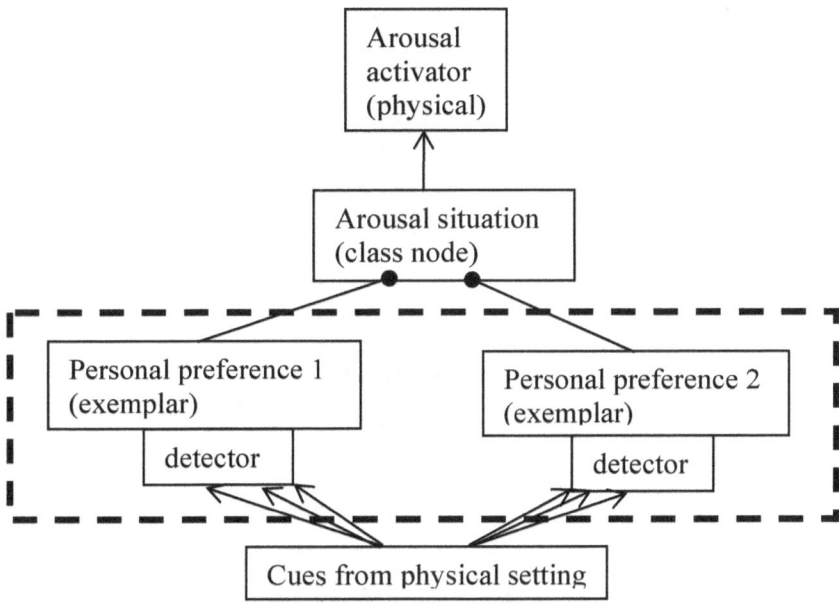

Figure 14 Sexual-arousal information-structure. Arrows indicate direction of activation flow. Circular arrows indicate exemplar-class relationships. Only two exemplars are illustrated. Dashed lines enclose concepts that are learned (nurture), the rest of the system is genetic (nature).

Establishing 'has-a' and 'is-a' relationships can be accomplished in several ways. One way is based on repetition. Patterns of nodes that fire repeatedly define an item, and become its parts. Patterns of firing nodes that are common to several items become cues to a class node, whose exemplars are the items that possess those common cues. Those mechanisms take part in building the sexual information structures.

Another important general learning mechanism is conditioning. It is suggested here that conditioning is a central mechanism that the brain uses in building the sexual preferences of individuals. Through conditioning, our brain learns the cues that would activate our sexual arousal and attraction in response to external stimuli, such as seeing a potential sexual partner.

In conditioning, a sensation or an emotion triggers the linking of an external stimulus with an internal response. For example, in Pavlov's famous experiment, the sensation of food triggers the linking of the sound of the bell to the salivation mechanism.

A big mystery is what triggers the linking of external stimuli to the physiological arousal and attractions mechanisms. What learning process result in an individual being aroused and attracted by firm breasts, by long legs, by a man, by a woman, by none-of-the-above, by all-of-the-above?

It has been suggested here that a certain change of emotions triggers the learning of sexual preferences. More specifically, the transition from feeling unsafe in the presence of another person to feeling safe with that person is the emotion that triggers the learning of the cues of sexual preferences. Behaviors that cause those feelings become sexual arousal cues. Such feelings were given the acronym SWAP, for safe with a partner or safe with another person. In order to arouse themselves, people act in ways that make them feel SWAP. In order to arouse their partner, people do things that make the partner feel SWAP. Numerous examples were brought to support that idea.

Activities and situations that have resulted in feeling SWAP are noticed by the brain, and their features are recruited to become cues that would trigger sexual arousal. That recruitment starts at early childhood and continues throughout life. The recruited cues include visual and other sensory features, such as body parts of people that created the feeling of SWAP in the child. Cues are also extracted from activities of the child himself or herself; for example, activities to please parents or other authoritative figures and activities to control those figures. Such activities are associated with the child feeling SWAP. Like with any information, those cues are generalized and expanded. They start as detectors of specific items and evolve into detectors of classes. For example, if a parent's hug created a feeling of SWAP, it is generalized to include a friendly hug from any person, and then to hugs that cause sexual arousal.

At a certain age, parts of the learned information become imprinted. They are fixated and they cannot be changed any more, except by adding new exemplars to the classes that have been fixated. The cues that would cause sexual arousal in the individual are there to stay. In the future, only situations that contain those features would arouse that person. The processes that develop the sexual arousal information structures, imprint them, and allow their expansion by adding new exemplars to the fixated classes are the same for hetero- and homosexual people.

The information structures of sexual attraction, which determine the sex of the target of the attraction, develop along the same principles as the arousal structures. However, here the trigger of the learning process must depend on the sexes of the affected individual and of the source of the information. The learning process itself must be universal; it should work in any society, anywhere in the world, and at any time throughout history. The trigger may be sensory or emotional, but it should be genetic. It is not known yet what that trigger is.

Voice may have a part in directing the learning of sexual attraction. The differences between men's and women's voices are genetic, and the way that voices are processed by the brain is also genetic. Therefore, voice satisfies the universality condition.

Some general behavioral traits that are influenced by sex hormones, such as aggressiveness, are statistically more prevalent in one sex than in the other. Those behaviors may also have a part in triggering the learning of sexual attraction.

Although scientists cannot identify yet the underlying biological processes of all aspects of sexual arousal and attraction, an extensive body of evidence suggests that the core of their information-structures is imprinted. It evolves unconsciously, starting at childhood, and is affected by non-sexual experiences.

Bob

Our sexual information structures, which determine our behavior, develop naturally, much like our other information structures. Yet, many people wish they could reprogram them. For example, some people wish that they were not gay; others wish that they could be aroused by things other than strange fetishes. What can you tell those people?

Morie

Like other information structures, sexual information structures have fixated parts and modifiable parts. If the source of an undesirable behavior is in a fixated part, it would be impossible to eradicate it. The most those people could hope for is to develop a bypass to the fixated nodes. However, even if such a bypass could be developed, it may cause mental stress. It will create conflicts with the fixated parts of the information structure.

As we have discussed earlier, there are situations that a bypass could not even be constructed. Nodes that have to be a part of the bypass cannot be accessed by the brain mechanisms that modify synaptic ties.

On the other hand, if the cause of the problem is not rooted in a fixated information structure, it could be modified. There might be activities or situations that would cause arousal and attraction, and at the same time would not cause adverse emotions. If replacement activities could be identified, brain learning mechanisms could add them as new exemplars to existing classes that cause arousal and attraction. They would activate the arousal and attraction centers without causing adverse reactions.

Of course, if the individual could embrace his or her own behavior, it would not be necessary to change it. In order to embrace that behavior, it should be associated with positive attributes that would counter the negative emotions with which it is associated originally.

Sol

We have said that there are no differences between the principles of operation of our sexual information structures and those of our other information structures. So, the same general conclusions that we have derived would apply also to expanding and modifying any information structure and the ways of thinking and behaviors that they bring about.

Morie

Yes.

FREE WILL

Claire

I have a small philosophical problem. I grew up to believe that I, like all other human beings, have free will. That there is some part in me that is not affected by any external factor, and that part enables me to make decisions that no external factor can affect. It is completely up to me to say yes or no to any option that I face.

However, if I follow the model that we have discussed, whether I would say yes or no to a given option depends on the information recorded in my brain in the form of synaptic weights. That information, in turn, has been recorded by definite mechanisms. The brain will retrieve the answer based on that recorded information, using retrieval mechanisms that are determined by my genes and my experiences. All that I have in my brain and how I use it is determined by my genes and by my experiences. Nothing that I do is by free will. Frankly, as far as having free will, there is no difference between me and a robot, albeit a sophisticated one.

Bob

What do you mean? If I ask you now, "will you marry me?" Are you saying that you don't have free will to decide?

Claire

I will treat it as a hypothetical proposal. You ask me and you do not know my answer. However, you know that I could say yes or no. Because of that, you assume that I have free will. But if you could map all the neurons in my brain, measure all their synaptic weights, and have a huge simulation program that can simulate on a computer the activity of each and every neuron in my brain, you would be able to know my answer, before I even open my mouth. The answer is already in my brain, and it has been pre-determined by my genes and my past experiences. I just retrieve it.

Bob

But then, if I find out that a "no" is coming my way, I could do and say all kinds of things before I ask you, in order to change your mind. Isn't that an indication that you have free will?

Claire

Not really. Your actions would alter some of the information recorded in my brain. If you then pose me the question, all those modifications would be included in the new simulation, and you could find out the revised answer, even before I know it. It is all pre-determined.

Bob

So, what is it that we have, if we really do not have free will?

Morie

We have awareness. We are aware of our situation, of our activity options, and of their possible consequences. We are also aware of what we actually do.

Our awareness enables us to concentrate on relevant parts of our information structure, before and after we act. We can concentrate on thoughts that originate from within us and thoughts that are triggered by outside entities such as family, friends and society in general. That concentration enables our brain to better respond to our surroundings. Awareness helps us to avoid making mistakes, to learn from our mistakes and to heed to advice of others.

Overall, our brain is a deterministic system. It consists of many elements whose cause-effect relationships are well established. Because of that, I must conclude that we do not have free will.

Bob

It sounds very pessimistic, because if I find myself in a tough spot, it does not matter what I do. Everything is pre-determined.

Morie

Not at all. All it means is that if you find yourself in a tough spot, your reaction is pre-determined. Whether you would fight it out, pray, seek help, give up and so on, is already in your system. Some super-wise entity would be able to predict your action. However, what you do may get you out of the tough situation or may not. It may make you happy, successful, or the opposite. Pre-determined does not mean that you should not do what you feel like doing. It does not mean that you are doomed. On the contrary, do what you have to do, and hope for the best.

It has been pre-determined that our meetings would come to an end. I hope that, like me, you have enjoyed our discussions, and that you put to good use the modifications of synaptic weights that they induced in your brains.

INDEX

www.ingramcontent.com/pod-product-compliance
Lightning Source LLC
Chambersburg PA
CBHW061354280526
45784CB00001B/254